D0455618

An Invented Life

An Invented Life

Reflections on Leadership and Change

Warren Bennis
Foreword by Tom Peters

▲▼ *Addison-Wesley Publishing Company*
Reading, Massachusetts · Menlo Park, California · New York
Don Mills, Ontario · Wokingham, England · Amsterdam · Bonn
Sydney · Singapore · Tokyo · Madrid · San Juan
Paris · Seoul · Milan · Mexico City · Taipei

HD
57.7
.B458
1993

All acknowledgments for permission to reprint previously published material can be found on pages 237–238.

Many of the designations used by manufacturers and sellers to distinguish their products are claimed as trademarks. Where those designations appear in this book and Addison-Wesley was aware of a trademark claim, the designations have been printed in initial capital letters (e.g., Rolodex).

Library of Congress Cataloging-in-Publication Data

Bennis, Warren G.
    An invented life : reflections on leadership and change / Warren Bennis : foreword by Tom Peters.
      p.   cm.
    Includes index.
    ISBN 0-201-63212-8
    1. Leadership.   2. Organizational change.   3. Business ethics.
I. Title.
HD57.7.B46   1993
658.4′09—dc20   92-35270
                                                    CIP

Copyright © 1993 by Warren Bennis Inc.
Foreword copyright © 1993 by Tom Peters

All rights reserved. No part of this publication may be reproduced, stored in a retrieval system, or transmitted, in any form or by any means, electronic, mechanical, photocopying, recording, or otherwise, without the prior written permission of the publisher. Printed in the United States of America. Published simultaneously in Canada.

Jacket design by One Plus One Studio
Text design by Diane Levy
Set in 11.5 point Century Old Style by Clarinda

1 2 3 4 5 6 7 8 9-MA-9796959493
First printing, February 1993

Addison-Wesley books are available at special discounts for bulk purchases by corporations, institutions, and other organizations. For more information, please contact:
Special Markets Department
Addison-Wesley Publishing Company
Reading, MA 01867
(617) 944–3700 x 2431

With Grace

CONCORDIA COLLEGE LIBRARY
2811 NE HOLMAN ST.
PORTLAND, OR 97211-6099

Contents |

Foreword |

Curiosity invariably gets the best of Warren Bennis. Lots of us are better off as a result. These pages are a wonderful tribute to a perpetually curious sixty-eight-year-old, who is as puzzled and thoughtful now as at age six or seven.

Bennis is a pioneer, and prescient. His work at MIT in the 1960s on group behavior foreshadowed—and helped bring about—today's headlong plunge into less hierarchical, more democratic and adaptive institutions, private and public. He's always burned his intellectual candle at both ends. While conducting meticulous research into the minutiae of social interaction, Warren was also risking self and soul as a leader of some of the first T-groups at the National Training Labs in Bethel, Maine. These intimate inner explorations, though, didn't stop him from engaging in the most oceanic prophecies. In a landmark 1964 *Harvard Business Review* article (see chapter 2), he and colleague Philip Slater astonishingly claimed that "democracy is inevitable." Only those of us who lived through McCarthyism and Khrushchev's shoe pounding at the United Nations can appreciate how outrageous such an idea sounded at the time. (The *Harvard Business Review* did Slater and Bennis the favor of reprinting the article in 1990; few of us have the opportunity of being able to say publicly, "I told you so.")

On the heels of that bold proclamation, just two years later, in 1966, came "The Coming Death of Bureaucracy" (chapter 4), in which Bennis insisted—bizarre, again, in the context of the times—that the centuries-old command-and-control, pyramidal organizational structures were doomed and that "adaptive, rapidly changing temporary systems" would be required to do tomorrow's work in the face of a society and economy beset by "chronic change." If only IBM, in whose *Think* magazine the article first appeared, had listened!

Warren's curiosity got the best of him again in 1967, when he began a decade of practicing what he had so long preached, first as provost at SUNY Buffalo and then as president of the University of Cincinnati. But to say he merely practiced what he preached is a gross understatement, for it turned out that leading in practice was a whole different kettle of fish from leading through influence via the printed page. The experience—containing at least as many potholes as peaks—tore Warren apart and doubtless contributed to his massive heart attack at age fifty-three.

With a stupendous career behind him, his place in the annals of social science ensured, scars up one side and down the other, inside and out, why bother with more? That damnable curiosity won out again, that's why. Bennis now turned his full attention and energy to figuring out where he'd been, and a series of methodically researched, landmark works on leadership ensued— with this, a personal reflection, the latest and in many senses the best.

Why best? Leadership, Bennis concluded while at the University of Cincinnati, is ultimately about the relationship between the leader-as-individual and the organization. In these pages he undertakes the sort of self-exploration and reflection that must be part and parcel of *any* successful leader's journey and makeup.

I love this book. It reflects humanity, openness, courage, and rigorous thinking in equal measures—and is a marvelous example (just the latest, in Warren's case) that good social science can be as scintillating and literate as it can rigorous.

I love this book, selfishly, because it confirms me in my perpetual confusion. On the one hand, Bennis has been constant in his concern with change, democracy, and bureaucracy for more than three decades. And yet he is as inconstant and fresh—and as timely and challenging—now as he was when I, as a neophyte MBA student in 1970, first came across his work ("The Coming

Death of Bureaucracy" was one of our first assigned readings at Stanford).

Bennis admits to having been mesmerized by his early mentors—Captain Bessinger, the high school dropout who was his company commander in 1944, when, at age nineteen, Bennis became the youngest infantry commander in the European theater of operations, and Doug McGregor, the social science giant whose thinking spurred much of Bennis's later work. I too am mesmerized by my mentors, and Warren is one of the most important. A reporter recently asked me what I wanted my epitaph to be. After declaring in no uncertain terms that I wasn't in any rush, I said, "He was curious to the end." Upon reading and reflecting on these pages, which came my way after that conversation, I know that I was, in a way, merely mouthing the words that are the signature of Warren Bennis's extraordinary career and life. Neither Bennis nor I is ready for epitaphs. But for those of you who wish to be inspired by the idea—and practice—of the curious life, richly spent, you'll do no better than to thoughtfully consider what follows.

Tom Peters

Preface |

I have been thinking about leadership almost as long as I have been thinking. My older brothers were my first teachers on this subject that would fascinate me for a lifetime. They were—and are—identical twins, alike in almost everything but their ability to lead. When we were growing up, one of my brothers was the archetypal natural leader, able to talk his teenage peers into doing things that parents never dreamed of, including ditching school for long periods of time. My other brother was the exact opposite, an innate follower without power, or even voice, within the group.

My brothers taught me the first two things I learned about leadership: that it is a function of character as well as behavior and that leading is better than following. As a child I began scrutinizing leaders, starting with my more charming, more authoritative brother, to discover how they controlled and shaped their organizations. I was hoping to learn their secrets: how to become a leader and, even more important, how leaders make things happen. Before I was out of my teens, I was studying leadership not in my living room or neighborhood but on the battlefields of Europe during World War II. Overnight I learned that a leader is not simply someone who experiences the personal exhilaration of being in charge. A leader is someone whose actions have the most profound consequences on other people's lives, for better or for worse, sometimes forever and ever.

This collection of essays, written over almost thirty years, contains the essence of what I have learned about leadership and change. It is an intensely personal book. It begins with an autobiographical sketch that attempts to link what I've experienced with what I've learned. Readers will discover how the war, Antioch College, graduate school in Cambridge, and other events and institutions shaped my ideas. They will learn what being a university president taught me about leadership and how shoe polish changed my life.

We are all children of our time, and my time—our time—has been one of the most remarkable in history. In reading these essays you will glimpse one revolution after another (the information explosion is only one), and you will see why I have come to believe that change is the only absolute.

A close reader will notice that some of the early essays contain now painful-to-behold references to "the leader . . . he." These were obviously written before the women's movement reminded us that language can oppress as well as liberate and that many of our best and brightest are women. You may also notice that I believe in recycling especially choice anecdotes.

Over the years I have given corporate executives and other leaders counsel on how best to lead, sharing such insights as the necessity of leading, not managing. But perhaps the most durable advice I can give leaders is to stay nimble. In this age of uncertainty, leaders must prepare for what has not yet been imagined. Leading today is like being a first-time parent—you have to do the right thing long before you fully understand the situation. Just one example: When I cowrote "Is Democracy Inevitable?" in 1964, no one imagined how profoundly Japan would shape the philosophy and practice of management, either domestically or globally. And yet the corporations from that era that survive are those which were flexible enough to respond quickly to the paradigm shift epitomized by the new Japan. Muhammad Ali was right: You have to flit like a butterfly, sting like a bee.

This book is part meditation, part how-to manual. I tend to share what I've learned in short bursts instead of extended analyses. Whenever possible I try to show rather than tell. Most of the substance of this book is contained in anecdotes. In this I wonder whether my teachers were not the deathless authors of the New

Testament, who routinely used parables to make their points. They realized people would learn more about unselfish compassion from the story of the Good Samaritan than from any long-winded treatise on altruism. I think there is too much pseudoscientific jargon written about organizations. We need more good stories.

Most of the essays in this book deal with facilitating leadership and managing change. But a significant minority deal with the ethics of organizational life. Effective leadership is not enough. It is essential that leaders in corporate and public life remember their societal obligations as well as their organizational ones. Every organization tempts its leaders to become preoccupied with the priorities of the moment at the cost of ignoring the over-arching questions that determine the quality of all our lives, such questions as: Is this right? Is it good for our children? Is it good for the planet? This book looks at some of the ethical dilemmas inherent in modern organizational life, including the often painful one of when it is better to resign than to collude.

Thirty years ago I was considered a futurist. Essentially that meant I had the chutzpah to predict what might happen, instead of limiting my analysis to what had already happened. This book ends with yet another stab at crystal gazing. It is a new essay that argues that World War II is finally over and that "federation is inevitable."

We can only hope.

Warren Bennis,
Santa Monica, 1993

1

An Invented Life: Shoe Polish, Milli Vanilli, and Sapiential Circles

I recently sat down and wrote this short autobiography at the request of a colleague who had solicited similar sketches from others in the field. Writing it was one of those heartening experiences, like a successful high school reunion, in which looking back made me not nostalgic for the past but grateful for the present. As I explain in the essay, this is a version of my life—a selection of the facts that tries to illuminate the work I've done that readers are most likely to know. It is more candid than exhaustive (always desirable when people are discussing themselves, I think). And it explains such mysteries as why I hate the accordion.

Not long after I sat down to write this brief intellectual autobiography, I had a small epiphany: I realized that what I was doing was actually *biography,* imagining a narrative about someone named myself. The result is a selection of stories, some called memories, that I—and to some extent others—have created to give coherence and meaning to my life.

What I'm talking about is self-invention. Imagination. That's basically how we get to know ourselves. People who cannot invent and reinvent themselves must be content with borrowed postures, secondhand ideas, fitting in instead of standing out. Inventing oneself is the opposite of accepting the roles we were brought up to play.

It's much like the distinction I made in my last book, *On Becoming a Leader,* between "once-borns" and "twice-borns."* The once-born's transition from home and family to independence is relatively easy. Twice-borns generally suffer as they grow up; they feel different, even isolated. Unsatisfied with life as it is, they write new lives for themselves. I'm one of those twice-born.

I believe in self-invention, have to believe in it, for reasons that will soon enough be clear. To be authentic is literally to be your own author (the words derive from the same Greek root), to discover your native energies and desires, and then to find your own way of acting on them. When you've done that, you are not existing simply to live up to an image posited by the culture, family tradition, or some other authority. When you write your own life, you have played the game that was natural for you to play. You have kept covenant with your own promise.

Shoe Polish

Samuel Beckett is my favorite playwright. Among the major writers of the twentieth century, he perhaps alone understood the relative insignificance of human existence in a vast, indifferent universe. His stage settings were virtually empty, an abyss without a timepiece, a space on the cusp of a precipice. Beckett dismissed the ordinary subject matter of the theater—social relations, struggles for power, and the like—as diversions masking the anguish and despair that are the essential human condition. Instead he asked bleak, existential questions: How do we come to terms with the fact that, without having asked for it, we have been thrown into being? And who are we? What does it mean when we say "I"?

In the 1958 play *Krapp's Last Tape,* Beckett continues his lifelong exploration of the mystery of self. An old man listens to the confessions he recorded in earlier years. To the old Krapp, the voice of the younger Krapp is sometimes that of a total stranger.

*I am indebted to Abraham Zaleznik for this idea.

In what sense, then, can the two Krapps be regarded as the same human being? In this essay I will try to bring the younger Bennis into some uneasy connection with the older one. But don't forget: I'm writing about a person who invented himself.

As I see it now, the landscape of my childhood was very like a Beckett stage set—barren, meager, endless. A little boy waited there for someone who might not, probably would not, show up. There were walk-ons occasionally: twin brothers ten years my senior, a father who worked eighteen hours a day (when he took off his shoes and soiled socks, the ring of dirt around his ankles had to be scrubbed off with a stiff-bristled brush), and a mother who liked vaudeville and played mah-jongg with her friends when she wasn't helping my father eke out an existence.

I was withdrawn, sullen, detached, removed from hope or desire, and probably depressed—"mopey," my father called it. I was also left pretty much alone. I had no close friends. I can't remember how I spent my time, except I know that I made up improved versions of my life that ran like twenty-four-hour newsreels in my mind.

I didn't much like school, and barely remember most of my teachers. Except for Miss Shirer. I liked Miss Shirer enormously. She taught the eighth grade, and she was almost famous because her older brother, William Shirer, was broadcasting from Berlin on CBS. I leaned into the radio whenever Shirer was on. That he was anti-Hitler was thrilling to a kid who, in 1938, often felt like the only Jew in Westwood, New Jersey, a town that richly deserved its reputation as a major stronghold of the German Bund.

On one psychically momentous occasion, Miss Shirer asked us to spend about ten minutes telling the class about our favorite hobby. I panicked. After all, I liked Miss Shirer a lot, but the truth was that I didn't have anything remotely like a favorite hobby. My efforts to develop recreational interests like those of the other guys had failed miserably. I was mediocre at sports. I was bored with stamps. I was too clumsy to tie dry flies, too nervous to hunt, too maladroit to build model airplanes out of balsa wood. What I finally decided to do, in a moment of desperate inspiration, was to bring a shoe box full of shoe polish, different colors and shades in cans and bottles, since the only palpable physical activity I regularly engaged in was shining the family shoes.

And so when it was my turn in the spotlight, I revealed the arcane nature of a new art form. I described in loving detail the nuances of my palette (I was especially good on the subtle differences between oxblood and maroon). I discoursed on the form and function of the various appliances needed to achieve an impressive tone and sheen. I argued both sides of the debate on solid versus liquid wax and wrapped it all up with a spirited disquisition on the multiple virtues of neat's-foot oil. It was a remarkable performance, if only because it was, from start to finish, an act of pure imagination. I could tell from her smile that Miss Shirer thought it was terrific. Even the class seemed impressed in a stupefied way. And there, in a flourish of brushes and shoe polish, a new Warren Bennis was born.

You should know a few other things about me before we draw any conclusions about my intellectual or academic contributions. My favorite essayist, Isaiah Berlin, once remarked that his reputation was based on systematic overestimation of his abilities. ("Long may this continue," he added merrily.) Over three decades I have written a great deal. A small portion of that work has had a life outside the pages of the journals in which it originally appeared. The most enduring examples are the work on planned change, the study of the stages of groups, the essay on the inevitability of democracy that almost miraculously came true twenty-six years after it was written, and the more recent work on leadership, particularly the ongoing analysis of why leaders can't lead. I am proud of that body of work, some written with distinguished colleagues. But there are moments when I look back and have my doubts. Where is the irrefutable masterpiece, the systematic application, the great theoretical treatise? When I think about the achievements of some of my peers, I sense a depth and continuity that is majestically alien to my own.

In an essay on Tolstoy, Berlin notes the distinction Tolstoy makes between foxes and hedgehogs. Foxes know many things of various degrees of importance, while hedgehogs know one big thing. Foxes are conceptualists. While the critics are fussing about the latest vulpine theory, the fox is already working on the next one. Hedgehogs, at their best, produce Darwins; at their worst, pedants. Foxes occasionally can claim an Einstein or an Oppenheimer but more often are simply dilettantes. I'm clearly a fox with a sneaking admiration for the hedgehog.

There's another aspect of my intellectual development that can't be dismissed: that is empathy and the role it has played in both my temperament and my work. I think the ability to read and respond to others has to do with my Jewishness and the sense of marginality that goes with it. Minorities have to be good at picking up subtle cues of rejection. Our moral radar is always switched on, ready to detect what is and what is not acceptable to the majority community. As Lionel Trilling once said, this enables us to understand the mind of the enemy. But Trilling also felt, as did Berlin and as do I, that empathy, or at least that part of it that involves eternal vigilance—the stethoscope always probing for danger—can also undermine one's critical abilities. There is what Berlin called the "fatal desire to please." Over the years I've come to value the ambivalent gift of empathy, but it did lead me, earlier on, to work on projects that a friend once described as "good boy" work—solid books of readings that cited everyone and his or her colleagues and left no idea unturned or fully developed.

Arthur Lovejoy, the historian of ideas, once wrote that every writer possesses, and often tries to hide, his or her distinctive "metaphysical pathos," those subterranean, often unconscious impulses and values that govern our choice of intellectual work. My way of putting it would be less metaphysical and abstract, although it makes the same point. It seems to me that the issues we select to study are almost always the underground churnings of unresolved conflicts—that our ideas stem from an attempt to solve our existential predicaments and that the unlikely force behind all rational problem solving is the need to quiet our demons. (We are all children of our time, and, needless to say, mine includes the golden age of psychoanalysis.)

These, then, are some of the early forces that shaped me:

• A family structured like a double helix, my brothers, bonded in unimaginable ways, in one strand, and a mother and father who rarely connected, in the other. I felt outside this structure, almost invisible, a nonparticipant observer. Growing up in a Jewish family in a gentile community, I rejected both. Talk about marginal.

• A search for power and potency born out of what I perceived as an unsuccessful father, who, prodded by my mother, moved from town to town, opening and soon closing a series of candy stores, malt shops, and soda stands. Like so many of my depression-era generation, I remember the day my father lost his last regular job as one of the most wretched and despondent of my life. Without realizing it then, I vowed never again to feel such utter hopelessness. Understandably, given the disappointments of his own working life, my father kept urging me to learn a trade, by which he meant carpentry or printing or tailoring. My mother, whose unrelenting forcefulness frightened me even more than my father's passivity, thought I should be a child movie star, on the order of Bobby Breen, who sang, on the brink of adolescence, like a castrated cantor. Recognizing that my voice was at best croaky and had a range of about half an octave, she insisted that I take ten accordion lessons from Pietro Agostino of Hackensack, New Jersey, sincerely believing that what my voice lacked, the 120 bass Hohner accordion could easily redeem. I did finally master "The Sharpshooters' March" and "Over the Waves." But even the professionally optimistic Pietro Agostino, who probably needed the $1.50 an hour I paid for my lesson and claimed to admire my "drive" (I schlepped the accordion on the Rockland County bus, ten miles from Westwood and back again, twice a week), felt that I lacked what he charitably described to my mother as "touch." Go figure.

• A terrible sense of uncertainty, which may be the human condition for non–grown-up humans. My only early certitudes were my ability to observe and an insatiable hunger to learn. The latter arose not out of anything as neutral as curiosity but because I needed the illusion of understanding in order to feel safe. After the shoe polish affair, I developed a growing sense of the power of the imagination, which may be the only real power children have.

I emerged from boyhood sure of only two things: that I never wanted to get on another bus carrying an accordion and that I didn't want to grow up to be like the people I already knew. It was almost time to invent a life of my own.

The U.S. Army: 1943–1947

The Army Specialized Training Program, better known as ASTP, beckoned. To qualify you had to pass a physical and demonstrate an IQ of 125 or so. Camp Hood, Texas (now Fort Hood), was the venue for its seventeen weeks of basic training. Following my stint at Camp Hood, I was shipped to UCLA for the collegiate portion of my army career. The army had no trouble matching my many incompetencies with a career track: I was assigned to major in "sanitary engineering." Fortunately, ASTP was dismantled in order to prepare for the D-day invasion of Normandy. And I was sent to Fort Benning, Georgia, to attend the infantry school there.

The so-called Benning School for Boys was the best-possible education for combat. If education is supposed to prepare you for what you will confront in real life, then the training there was near perfect. German villages and cities were reproduced on the base, and we were drilled in what we were likely to encounter as the Germans desperately resisted the end of the war. While it's fair to say no one is every truly prepared for combat, Fort Benning came pretty damned close.

In 1944 I was commissioned as a second lieutenant and sent almost immediately to the European theater of operations, first as a platoon commander and later as a company commander. I was nineteen. (Later I learned I had been the youngest infantry officer in the ETO.) The company I joined had been savaged during the Battle of the Bulge. Out of 189 men and six officers, the normal size of an infantry company, only 60 men and two officers remained. One of the two, Claude Williams, had graduated from Benning just two months before I did. The other, the CO, Captain Bessinger, had in civilian life been the caretaker of the Vanderbilt mansion in Asheville, North Carolina. He was old, I thought, almost thirty-five, and half deaf because of the incessant roar of German antitank guns. He was also one of the finest leaders I've ever met.

As noted earlier, I had first become interested in leadership watching my twin brothers, one of whom effortlessly initiated activities that attracted other kids (including a rather tame teenage gang), while the other couldn't influence his way into a stickball game. My army experience affirmed my lifelong interest in the topic.

In the army I saw firsthand the consequences of good and bad leadership in the simplest and starkest terms—morale, tank support that would or would not be where it was supposed to be, wounds, body counts. The army was the first organization I was to observe close-up and in-depth. And although I have been in pleasanter classrooms, it was an excellent place to study such organizational realities as the effects of command-and-control leadership and the paralyzing impact of institutional bureaucracy.

Captain Bessinger was a wonder. Despite his poor hearing, he really listened to the men, inspired them, and protected them from the whims of the brass. In every way Bessinger embodied what Doug McGregor, my mentor later in college, immortalized as a "Theory Y" orientation. Bessinger was also my first role model, although I didn't know that phrase then and even now the banality of the term doesn't do justice to the man. A high school dropout, Bessinger literally kept me alive. He taught me how to identify different kinds of German artillery by their sound. He taught me how and when to duck. He also had a quality I deeply respect but have never been able to emulate—the courage to be patient.

After a month or so in combat, I became weakly confident that I wasn't going to bolt or go nuts (I was less sure I wasn't going to die). And in the time-honored army fashion, I began grumbling about the conditions we were fighting in. We had inadequate air cover and tank support, incompetent "forward observers" from the artillery, delays in getting reserves, unspeakable rations, and so on. Each day my voice would grow more strident, and each day Captain Bessinger would chew his tobacco and listen, with less and less of his legendary patience. One day (only a few weeks before the war ended, as it happened) I blurted out, "I, for one, don't know how the hell we're going to win this f——ing war unless. . . ." Finally, the captain spat out his plug of Red Man, looked at me through sad, beagle eyes, and said, "Shit, kid, they've got an army too."

Bessinger had given me exactly the useful truth I needed. The Germans did have an army, an army composed largely of hungry fourteen- and fifteen-year-old kids, shooting wooden bullets because they had run out of metal casings (the wooden bullets exploded hideously on contact). They were even more frightened than I was and had to contend with a bureaucracy that was at least as bad.

I wonder how much that story conveys to someone who wasn't there. The army in wartime was an organization, unlike most I've studied since, in which miscommunications and errors in judgment could kill you. I was a teenager, desperately trying to make sufficient sense of the general chaos to stay alive, and along came a person who listened, as empathically as possible, and who, despite coming from a totally different culture from my own, was able to transcend age, rank, and ethnic and religious background to help me cope with our mutual dilemma. Bessinger, whom I haven't seen since he was wounded in 1945, wasn't just a leader; he was the kind of leader you read about in the Bible.

I came very close to signing up for a career in the army once the war ended. My division, the Sixty-third, was dismantled, and I was eventually sent to European headquarters in Frankfurt, where I served in the transportation corps. I had a Jeep and a driver, an apartment in the Frankfurt compound, and membership in the officer's club. I was twenty. The army had already served me well, if only by giving me an honorable way to leave my family. It had taught me self-reliance and the extraordinary power that comes of being organized and using your time efficiently. Frankfurt was a kind of finishing school. I learned what fork to use and how dry a martini should be. I took weekend trips to Luxembourg City, Wengen, Switzerland, and Bad Homburg with a sweet "older woman" (she was going on twenty-six), a captain in the WACS. Among the heady things she taught me was how to eat an artichoke.

The main reason I didn't sign up for an army career was that my runner, Gunnar, had beguiled me with stories about the college he had been attending before the war. He loved the school, located in bucolic-sounding Yellow Springs, Ohio, in large measure because it allowed him to work part of the year and attend classes the rest of the time. The college had a strange name: Antioch.

Gunnar told me he wanted to become a clinical psychologist, and he had already worked for the Psychological Corporation, validating tests, and in the personnel department of Macy's, testing job applicants. He said that the courses he had taken had prepared him for the jobs and that the jobs had enhanced his classroom experience. I was fascinated by Gunnar's tales of Antioch. No one in my entire family had gone to college—no one. But Antioch intrigued me, and I figured I could afford it, given both the GI Bill and the co-op job system.

Gunnar was killed by an errant canister of white phosphorus on the last day of fighting in the town of Budesheim, Germany. And I, after serving two more years in Europe, was accepted as a freshman at Antioch.

Antioch College: 1947–1951

I took the train to Antioch. (It actually stopped at Springfield, and I hitchhiked the last ten miles to Yellow Springs.) As we neared my destination, my seatmate couldn't resist asking me why I wanted to go to a "Commie school," with its "nigger-lovers, pinkos, and people who believe in free love." (What, I have wondered ever since, is the opposite of free love—expensive love?)

Antioch was progressive. Even then we called it "politically correct," although without the ironic tone we use today. In many ways it was an ideal community. The campus heroes were intellectuals. There were no Greek societies or social clubs. People of color were celebrated, the Young Communist League and followers of Henry Wallace were taken seriously, and the talk was ferocious, utopian, and unending.

But for all its commitment to diversity and independent thought, Antioch had a definite subculture, an unwritten Antioch way. We ordered our organically grown wheat from Deaf (pronounced Deef) County, Texas; our Telemann from Sam Goody in New York; and Dwight MacDonald's *Politics* and *The Nation* from Greenwich Village. The books we read were Erich Fromm's *Escape from Freedom,* Bertrand Wolfe's *Three Who Made a Revolution,* Edmund Wilson's *To the Finland Station* and *Axel's Castle,* Marquis Child's *Sweden: The Third Way,* Djuna Barnes's *Nightwood,* Malcolm Lowrey's *Under the Volcano* (we all knew that it was his one and only novel and that he had died too young from booze), T. S. Eliot's *The Cocktail Party,* and the complete works of Thomas Mann, Hemingway, Fitzgerald, Dos Passos, Ford Madox Ford, and Virginia Woolf. We uniformly vilified the literary upstarts: Norman Mailer, Irwin Shaw, Herman Wouk, and James Jones.

The army taught me the value of being organized. At Antioch I learned to have opinions. That may not sound very important, but it amounted to a personal paradigm shift. Before college, I had been like Olenka in Chekhov's story "The Darling." Olenka, Chekhov writes, "saw the objects about her and understood what was going on, but she could not form an opinion about anything and did not know what to talk about. You see, for instance, a bottle, or the rain, or a peasant driving his cart, but what the bottle is for, or the rain, or the peasant, and what is the meaning of it, you can't say, and could not even for a thousand rubles."

What freedom, what liberation, to have opinions, sometimes based on reason and evidence, sometimes based on nothing more than the liberal campus zeitgeist. There were times at Antioch when a particular politically correct opinion would run through the entire population like a flu epidemic (vegetarianism and the superiority of home weaving were two I recall). But all the same, having opinions was, at least for me, tantamount to developing a personal identity.

Later on, as I became a nimbler, more seasoned Antiochian, I developed a whole set of counteropinions. I began tweaking, sometimes reviling, the campus's more doctrinaire positions in a series of pseudonymous satires in the college literary magazine, writing under the name Dr. Gruppen Ausgefundener (Dr. Group Finder-Outer). The satires caused George Geiger, our venerable professor of philosophy, to compare me with S. J. Perelman (what a coup!) and made just about everybody look at me differently.

As a result, I was "tapped" (informally, of course) by the campus intellectuals, despite my philistine interest in social psychology and economics. The cognoscenti were easy to spot at Antioch. They wore army fatigues (the women especially); smoked Camels in long black cigarette holders; drank beer at night at Com's, a black bar in town; listened with closed eyes to Johnny Coltrane keening on the jukebox; played esoteric games like Botticelli; regularly threatened to transfer to Columbia or Chicago; and constantly said "fug," emulating James Jones's queer contraction in *From Here to Eternity,* a book everyone claimed to loathe. I trembled with delight when I was invited to join them at Com's.

Henry Broude was one of the Brahmins, then a senior (I was a junior, or, as we, finding all labels of rank offensive, said, "a third-year student"). He was Waldemar Carlson's teaching assistant in Fiscal Policy, which introduced me to Keynes. Henry, now a distinguished economist at Yale, probably doesn't remember this, but after he read my term paper for the course, he said I might want to consider graduate school and mentioned in particular Harvard or MIT. The Nobel laureate John Franck once said that he always knew when he had heard a good idea because of the feeling of terror that seized him. I was seized.

There were at least three other things that influenced me during those Antioch years. First was the famous co-op program, pioneered by one of the college's great presidents, Arthur Morgan, an engineer who was Roosevelt's first head of the Tennessee Valley Authority. Antioch's program was mandatory, complex, and extraordinary. You came to campus for eight weeks; then, with the counsel of an adviser, decided on a job somewhere in the world—usually a fairly large urban center—and worked for twelve weeks; then returned to campus for sixteen weeks. It normally took five years to finish Antioch decades before that became the national norm. I did it in four because, in classic Antioch fashion, I got co-op credit for my four years in the army.

Splicing classroom experience with real-world work was a wonderful way to explore the relationship or lack thereof between theory and practice, word and act, those who make history and those who study it. There was an exquisite tension between the idealistic tilting at windmills that went on on campus and the inevitable compromises of the workplace.

Second, Antioch forced me to confront, for the first of innumerable times, both my desire to achieve personal satisfaction and the often conflicting urge to live up to the motto of Antioch's founding president, Horace Mann: "Be ashamed to die until you have won some victory for humanity." That tension between self-expression and civic responsibility continues to trouble me, perhaps even more now than it did then.

Third, Antioch deserves credit for teaching me to beware of totalized explanations of life and other mysteries. Despite the campus's sometimes unfortunate tendency toward groupthink, there were so many competing ideas poking at you that you couldn't help developing a healthy skepticism about commissars of thought. The quest for a Parnassian truth, a rule or rules for everything, was what my heart wanted but my mind rejected. The Spanish film director Luis Buñuel used to say, "I would give my life for a man who is looking for the truth, but I would gladly kill a man who thinks he has found the truth." Although there were a few "true believers" on the faculty, most of the professors were skeptics with little patience for universal systems. Certainly the great lesson of the first half of the twentieth century has been that overarching systems or theories that eliminate the opportunity for independent thought lead to totalitarianism. We had experienced the savagery of Nazism, the horrors of Stalinism, the limits of Marxism and Freudianism. We were sufficiently adolescent to continue to seek the grail of grails. But we also knew, in our disappointed hearts, that the Truth could be fatal to millions.

But my most important influence at Antioch by far was its president, Doug McGregor. He came there in 1948 at the age of forty-two, open, broad-grinned, and tweedy from MIT, where he had started an industrial psychology department. I don't think there was a search committee in those days, or else Arthur Morgan simply ignored it. Morgan visited McGregor at MIT, liked him enormously, and asked him to become college president.

Doug was at Antioch for six years and turned the school on its head. At his very first assembly, he announced, while our collective jaws dropped, that he valued his four years in analysis more than his four years as an undergraduate, that he hadn't the faintest idea what the students or faculty wanted, and that maybe the campus should shut down for a week while we had some "goal discussions" in small groups. Soon after, goal discussions were initiated (much to the consternation of the Brahmins, who thought they amounted to "pooling ignorance"). Those sessions redefined our collective aspirations, focused our vision for our education, and constituted a superb example of how change is facilitated by involving those who will be most affected.

In a foreword to a book of Doug's essays, published after his death at the tragically early age of fifty-nine, I described Doug as "a born innovator, a born experimenter. He refused to accept what was, or the traditional, uncritically, and it may be that his greatest and most permanent achievement was to create an atmosphere in which students, as well as faculty, were stimulated to question and challenge continually in an effort to create an educational program that had a relationship to the whole life of the individual. . . . If there was anything he was trying to overcome or destroy, it was the institutional habit of talking about the virtue of democracy while running affairs autocratically."

By the time Doug returned to MIT in 1954 to start a new program in organizational studies, he had already laid the groundwork for what today is called organizational behavior, human relations, or personnel management. As Mason Haire, one of his colleagues at MIT in the 1940s, pointed out, Doug created much of the professional field in which he operated: "Much of the work that goes on now couldn't have happened if he had never been."

If Captain Bessinger saved my life, Doug McGregor surely shaped it. In my final year at Antioch, I took a tutorial with him on "superior-subordinate relationships and leadership" and several courses on group dynamics, taught by his MIT colleague, the maverick psychologist Irving Knickerbocker.

I liked everything about Doug and tried to be like him in every conceivable way. I started smoking a pipe, tried to dress the way he did (though I was a 38 short and he was at least a 42 long), and applied to MIT for graduate work, although I had little idea what it would entail or even that it would funnel and focus my later choices. The truth is that I wouldn't have gotten into MIT without his recommendation; nor would I have gotten tenure there without his full-throated endorsement (he threatened to quit if it wasn't granted unanimously); nor would I have sought a university presidency; nor, in short, would my life have taken the direction it has.

In a recent interview a British journalist, David Oates, talks about my dissociation of myself from my family and quotes me as saying, "I was brazen in getting teachers at school to make me the favorite son. I kept being adopted by intellectual father figures and was shameless at sucking up to mentors." Sounds awful, doesn't it? The phrase "sucking up" is appalling, I know. Without repudiating that confession, I'd like to reframe it: I did cultivate major figures in my field—Doug McGregor, Carl Rogers, Abe Maslow, Erik Erikson, Peter Drucker, and others. While I was not unmindful of what their patronage might mean professionally, the truth is that I couldn't resist the power of their ideas and their personalities. I was so drawn to genius, perhaps, because I felt so ordinary myself.

MIT and Milli Vanilli: 1951–1956

MIT was as different from Antioch as Cambridge from Yellow Springs. My straight-A performance at Antioch and Doug's three-page letter of recommendation were the sole reasons I was admitted to the MIT Economics Department. In college I had taken algebra and introductory physics, and that was about it. Most of my new MIT classmates had taken advanced calculus, knew at least the rudiments of set theory, understood Markov chains, and had mastered Boolean algebra. At my first interview, the then–admissions officer, economist Charles P. Kindleberger, outlined the courses I'd have to take to catch up and confessed, "We didn't exactly throw our hats in the air when we saw your application." It was a daunting revelation.

MIT was a confusing cocktail of makeup mathematics; philosophy of science; microeconomics from a great teacher, Bob Bishop; more economics from a great mind, Paul Samuelson; industrial statistics from Bob Solow; consumer economics from Franco Modigliani; and economic history from Walt Rostow (my only honest A). A pop quiz: Three of the above are Nobel laureates. Who are they?

But there were also George Shultz, who taught labor economics, and Alex Bavelas, who was the most brilliant designer of small-group experiments who ever lived. And then there was another early mentor, the most playful professor during my MIT days, Herb Shepard, who taught me that groups are real, even if they don't have spinal cords, and introduced me to Harry Stack Sullivan, Karen Horney, Erving Goffman, Norbert Wiener, L. J. Henderson, Elton Mayo, Walter Cannon, and a raft of people who were developing networks—not the human networks I was familiar with but electronic ones. In other words, a raft of people who were inventing the present day.

What a dazzling group! I was sometimes befuddled, routinely intimidated, and thoroughly outclassed by these new colleagues of mine, including my fellow graduate students (there were no women in the department then, and the only minority group was Canadians). It is probably true that I was the least prepared, the least mathematically inclined (Samuelson put it charitably when he said I lacked "mathematical flair"), and the only one who really wondered why he was there. The painful truth is that in Samuelson's seminar, whenever he was summing up, saying something like, "Well, that's the theory of duopoly," he would look over at me and ask, "Warren, are you with me?" If I nodded yes, he knew he was free to go on to the next unfathomable point (perhaps the Stackleburg point!), certain that everyone else in the class had got it hours before.

To get through the Ph.D. program, a sometimes uneasy amalgam of economic theory and social science, I began to memorize and mimic. I don't think I really understood what the Walrasian General Theory of Equilibrium was all about, but I was perfectly capable of memorizing the equations. I imitated my professors and the brightest of my fellow graduate students. For roughly two years I lip-synced what I heard, Milli Vanilli style. Eventually the words I formed on my lips came more naturally, but I often wondered whether I was kidding myself and should try my hand at something else (tailoring, Dad?).

Perhaps most of us learn through a form of lip-syncing, but I often found the process terribly confusing. Sometimes I would identify with Herb Shepard, cuddly, empathic, and warm, and at other times with Bob Solow, Brooklyn-tough, caustic, and wonderfully lucid. Sometimes my model was Talcott Parsons, who taught sociology at Harvard (I took up to half my social science courses at Harvard under an agreement between MIT and Harvard that may still be in effect today) and was unfathomable in ways that Samuelson couldn't dream of. Parsons was a Weberian sociologist prone to neologisms whom the graduate students dubbed "Talk-a-Lot" Parsons in our Christmas play. Sometimes my model was Alex Bavelas, with his Gretsky-like touch in setting up small-group experiments, so subtle and brilliant and utterly charming.

I lip-synced all of them. When I began teaching undergraduates, I didn't always know who I'd be that day or what I would sound like. On some days I thought of myself as a total fraud. Especially the day I bought an expensive, nonrequired book on microeconomics at the Harvard Coop after Paul Samuelson had casually mentioned it in class. Written by Stanford professor Tibor Scitovsky, it was called *Competition among the Few.* It cost $5.95 (the rent on my small apartment on Gray Street was only $9.00 a week), and when I handed the clerk my Coop credit card, I became woozy with self-doubt. What was I doing, lip-syncing idols and going broke buying books I could hardly understand?

After completing my thesis in less than a year and a half—something of a miracle, given my state of mind—I taught social psychology for one year (1955–1956) as an assistant professor and then left because the department had a policy against hiring its own until they had taught elsewhere for at least five years.

Actually, I was glad to leave. I still couldn't pass Paul Samuelson in the corridor without stammering over his first name, and I had begun to suspect that the MIT approach to truth, mathematical and quantitative, was not only beyond me but limited as well. Logical positivism had stormed the social sciences in the 1950s with its belief that all certifiable truths about human behavior could be predicted with scientific certainty. I wasn't sure about that then and am even more dubious about it today. However meticulously obtained, facts are rarely unassailable. And I was tired of fighting my natural impulse, revealed as long ago as Miss Shirer's eighth-grade class, to poeticize the materials at hand and give them a distinctive shape.

The thing I feared most, even beyond incompetence (which I thought about constantly), was that I would become an anemic heir to the majestic but alien minds of my teachers. And I was terribly tired of moving my lips to someone else's tune.

Bethel, Boston, and Sapiential Circles: 1955–1967

Life picked up steam that last year at MIT. It's a jumble of memories now, but a nascent career seemed to be taking shape. A career as what was less obvious. While officially my degree was in economics, I knew in my heart I was a generalist. At Cambridge dinner parties, where one's discipline was an identity card, I sometimes blushed when I described myself as an economist. Often I would simply say that I taught at MIT and a respectful hush would fall over the group, as if those letters sufficed for station identification.

The lack of a clear-cut professional identity had its advantages. I was an inkblot on which others could project their needs. Once, for instance, the editor of a mildly radical journal published at

Brandeis called and asked if I would join the editorial board. I did so gladly. Later the editor, Lew Coser, told me that the key reason for the invitation was that the journal needed a gentile on its masthead and thought one from MIT would look especially good. Actually, I enjoyed the editing involved, but, more important, our editorial board meetings put me in touch with a group of scholars more interested in ideas than in their measurement—people like Coser and his wife, Rose; Kurt Wolff; and Maury Stein. My Brandeis circle thought of me as an economist, whereas at MIT I was generally regarded as a social psychologist. Not having to be pinned down was fine with me.

In 1955 I was invited to Bethel, Maine, the summer headquarters of the National Training Laboratories. At Bethel, as we all called it, everyone was buzzing about a new social invention called T-groups (the T stood for training). Established in 1947 by the redoubtable refugee psychologist Kurt Lewin, NTL crackled with intellectual energy and the heady sense that some major discovery about the real nature of groups was taking place.

Bethel was singularly fortunate in having three genuine social revolutionaries on hand to help it weather Lewin's tragically premature death at the age of forty-seven. Ronald Lippitt, Kenneth Benne, and Leland Bradford each brought their own special gifts to bear. A distinguished young social psychologist, Ron Lippitt contributed intellectual rigor and methodological sophistication. No one could articulate the extraordinary spirit of the place better than Ken Benne, Bethel's resident philosopher. He was a dazzling intellectual "fox" who has taught me much. And holding it all together was Lee Bradford, who was both a visionary and a first-rate manager, a man who continued to dream even as he kept an institution full of dreamers running smoothly.

Leading a T-group at Bethel was a wild, exhilarating experience—"a trip," as the nation would begin to say a decade later. For a period of two weeks a group of strangers was brought together and asked to leave behind the roles, constraints, and norms of everyday life. People screamed, people guffawed, people wept, people talked: You never knew what would happen next. In the micro-utopias of Bethel, I discovered what life could be like when the usual mechanisms that govern our quotidian lives are absent. As the group developed and evolved, I saw how we search for structure and support and how we recoil from some individuals and align ourselves with others. I also realized how deeply our attitudes toward authority are buried and how stubborn they are.

In the supercharged atmosphere I was sometimes over-whelmed by what I saw and felt. When a group really came to-gether, when the communication was free and telling and truth-ful, you could practically feel the bonds between us expanding and deepening like the intertwined roots of enormous trees. Once in a while I felt that our bodies were somehow actually joined, like those of Siamese twins, so that the emotions circulat-ed between the members of the group, creating some superior new social organism. It was heady stuff that made what took place in the typical academic small-group lab look as drab as a black-and-white movie.

Later that year Herb Shepard and I wrote two articles on "natu-ral groups" in which we described the stages of group growth that I had experienced so vividly at Bethel. The articles were published back to back—an unusual move—in *Human Relations,* then the most prestigious journal in social psychology. No men-tion was made, needless to say, of merging roots or Siamese twins.

Although I was unaware of it at the time, it seems clear to me now that the study of group dynamics that flourished after World War II was as much a response to the recent past as a leap into the future. And in large part, I now believe, it was a reaction to Hitler.

At least two of the giants in the field, Kurt Lewin and Fritz Redl, were Jewish refugees from Hitler's reign of terror. It seems almost inevitable that they would have developed a profound faith in democratic groups, given their firsthand experience of the de-structive power of charismatic leaders, including their ability to enslave their followers. Lewin's early experiments, as well as those of his students, seem now almost foreordained to demon-strate that democratic groups are not only more fulfilled psychologically but more efficient, especially under complex and changing conditions. The theory that democratic groups are superior struck a sympathetic chord with a whole generation that had heard the ominous roar of a nondemocratic group cheering Hitler at Nuremberg.

Interestingly, those Americans who were drawn to Kurt Lewin's theories were almost all midwestern populists with a homegrown antiauthoritarian tradition of their own. I'm referring to Bradford, Benne, and Lippitt, of course, but also to Herb Thelen, Ren Likert, Doug McGregor, and many others. In that postwar period social scientists and related researchers, myself included, tended to view all authority with deep-seated skepticism, if not suspicion. We adopted much the same stance toward leaders that Baudelaire took toward newspapers: that you might learn from them if you read them with the proper contempt.

The three years between the time I left MIT and returned there in 1959 were spent teaching at Boston University. I worked mainly with Ken Benne and Bob Chin, as well as with the head of BU's Psychology Department, Nathan Maccoby. I also taught at Harvard with Freed Bales, Phil Slater, and Ted Mills and did research on groups with Will Schutz and Tim Leary (both at Harvard in the antediluvian age before Esalen and LSD, respectively) and with the psychoanalyst Elvin Semrad at what was irreverently called Boston Psycho.

At that time I was also undergoing six years of psychoanalysis with a Boston-trained analyst. That means, for those of you uninitiated in the therapeutic and intellectual folklore of the 1950s, a real, orthodox Freudian analysis. My analyst changed the pillow case for each analysand, every fifty-minute hour.

That was a rich, tumultuous, enchanting time in my life. Only my mother seemed unwowed by all I was experiencing. I remember telling her in 1959 that I had been psychoanalyzed. I sensed that she didn't like the thought of me untangling my psyche in the presence of a non-Bennis, but her only response was to ask how much I had paid the doctor. I told her—$3.00 an hour for the first three years, $15.00 an hour after that. "And you went how often?" she asked. I told her—five days a week for the first three years and four days a week for the next three. She was silent for several minutes and then said, "Hmmm . . . that comes to quite a bit of money." Another pause, and she said, "Son, I wish you had taken that money and spent it on yourself."

In 1959 Doug McGregor invited me to return to MIT, where he was heading up a department in the new Sloan School of Management. He had already recruited a formidable team: Don Marquis and Ed Schein, in particular, and later Bill Evan, Per Soelberg, Tom Allen, Mason Haire, Harry Levinson, Bob Kahn, Dave Berlew, and Fritz Steele.

Those years at BU and MIT were my best in academic terms. My output was prodigious—everything from tightly designed experiments in small-group communications to psychoanalytic exegeses of the schism between C. P. Snow's "Two Cultures." With Benne and Chin I produced my first book, a selection of readings titled *The Planning of Change* (1961). I also did another book of readings in interpersonal dynamics, in collaboration with Ed Schein, Dave Berlew, and Fritz Steele.

Both books eventually went into four editions, but even more thrilling than their success was the experience of working as part of an intellectual team. Sometimes when a group of talented people comes together, even if only for a short time, something wonderful happens. Each individual energizes the others, teaches them and learns as well. When everything goes right, this creative collaboration produces something new and important. Twenty-five years ago I mentioned this phenomenon—which has led to such diverse achievements as the Bauhaus School and the atom bomb—to Margaret Mead, who was giving a speech at Harvard. I asked her whether much had been written about it. She said no. "You write about it," she said. "And call the book *Sapiential Circles*." And that, finally, is what I am doing now.

SUNY Buffalo: 1967–1971

I made the pilgrimage to Buffalo, as did many others, largely under the spell of Martin Meyerson's bold dreams and blandishments. I visited the western New York campus, and later, in my

Beacon Hill home in Boston, I asked Martin what his own goal for Buffalo was. Always thoughtful, he hesitated a moment and said, "To make it a university where I would like to stay and be a professor after finishing my administrative responsibilities."

Over the next six weeks I spent most of my time trying to decide whether or not I should leave Boston for Meyerson's grand experiment on the shores of Lake Erie. It was an excruciating period, made even more tense by the fact that my wife had had a miscarriage after her first trip to Buffalo and had to spend most of the time in bed. My memory of that late-winter trip to Buffalo is one of exhaustion. And God, was it cold! As I walked down the roll-away stairs into the white swirl of the runway at Buffalo International Airport, all I could think of was someone's bitchy observation that summer in Buffalo was three weeks of bad iceskating. (Actually, summers there are lovely—as summers are only in places where you fully understand the alternative.)

My colleagues at the Sloan School were unsympathetic when I announced I was considering an offer to be provost at the State University of New York at Buffalo. Their attitude toward administrative jobs at any university bordered on contempt. "God," one asked, "why do you want to spend your time shuffling papers?" And then, of course, there were the snowblower jokes. It was difficult to explain to them what I was going through, but now I realize it was a genuine crisis. I was haunted, almost obsessed, by the need not just to teach and research management and leadership but to experience it firsthand. I wanted, as Shakespeare put it, to know "a hawk from a handsaw."

I turned to everyone I could think of for advice. I remember calling David Riesman at Harvard one day early in March 1967. He had taught at the law school at Buffalo when it was still a private university and was advising Meyerson on social science matters. David gave me a realistic assessment of the academic state of the university (many mediocre departments; some first-rate ones, including the quirky, creative English Department). He also put in a perversely good word for the industrial landscape, with its "chartreuse and black and mauve smoke against the steel-gray sky."

Still perplexed, I approached a friend at the Harvard Business School who studied mathematical models of decision making. He said he had once been in a similar position and had gone to his dean for advice. "Why don't you work it out mathematically?" the dean asked him. My friend howled when he recalled his indignant response: "But this is important!"

Meyerson kept up his dignified campaign of persuasion. His great gift as a recruiter was his ability to transmogrify all the highly visible drawbacks of Buffalo and make them reappear in the guise of exhilarating challenges. What a pleasure it was to be with him! Meyerson has a wonderfully agile, broad-ranging mind; he can think of nine things at once. Moreover, he seemed to know everybody. (It's from him that I learned the importance of the Rolodex.)

In recruiting, Meyerson's ace in the hole was a truly monumental vision for transforming Buffalo from a conventional university to an academic New Jerusalem—"the Berkeley of the East," as he liked to say. The ideas were stunning: decentralization of authority; dozens of new colleges that would function as intimate "intellectual neighborhoods"; universitywide research centers to grapple with urban studies, higher education, and other major issues; a new campus to be built from the ground up—the list went on and on. It was a seductive dream that tended to drive out any trivial-seeming qualms I had about the weather and the number of good bookstores in Buffalo compared with Boston.

While I was pondering the Buffalo offer, I was also considering two others—one from the Salk Institute and the other, ironically, to be vice-president for academic affairs at USC. As I weighed and reweighed alternatives, real life intruded one morning when I discovered that someone had broken into our house and stolen the family's winter coats from the vestibule.

The Boston winter ended that day. Drawn outside by the sweet air of early spring, I took a brisk, pleasant stroll over to Filene's Basement, where I bought a new overcoat—the heaviest alpaca snowcoat in the store.

Obviously, without realizing it I had made up my mind.

Two other provost candidates handpicked by Meyerson accepted about the same time. Eric Larrabee, a suave and brilliant New York editor *(Harper's, Horizon),* became provost of arts and letters, and Karl Willenbrock agreed to leave his post as an associate dean at Harvard to head Buffalo's engineering faculty. Four other provosts were recruited from the existing faculty.

Larrabee, Willenbrock, and I—Meyerson's chief outside recruits—arrived on campus that fall with all the optimism and confidence of young princes joining a crusade. Though only Willenbrock was a seasoned administrator, our relative inexperience didn't deter us in the least. We were sure that in this academic Great Good Place, creativity would count for more than traditional training and ordinary credentials. Meyerson had emphasized this point repeatedly. If in fact his idea of unorthodox training was a degree from Harvard in a field other than the one you were appointed to teach in, then so be it. We had no doubt we could set Buffalo free.

We certainly looked the part. Buffalo is a town where you don't have to make fashion statements. Swathed in down, everybody looks pretty much like the Michelin man for most of the winter. But we three had style, even panache. I remember the entrance we made at the first provosts' meeting. Karl invited me to have breakfast with him before the meeting at the Frank Lloyd Wright–designed house he had picked up for a song on his arrival in Buffalo. Afterward he and I climbed into his white Porsche convertible (a sure sign of an out-of-towner—people in upstate New York tend to favor cars with front-wheel drive in colors that will stand out against the snow, not disappear in it). As we were about to drive away, Karl's wife, Millie, came out to say good-bye and handed us berets to protect us from the wind.

We drove from Willenbrock's Frank Lloyd Wright house to Larrabee's Frank Lloyd Wright house (which was adjacent to Meyerson's Frank Lloyd Wright house). Eric was waiting for us, a homburg perched on his aristocratic head, his umbrella furled. I remember the sense of euphoria I had as we drove toward campus in the warm September sun and my complete confidence that, if nothing else, we epitomized Buffalo's new look. God only knows what the natives thought as we drove onto the tree-lined campus. To some, I'm sure, we must have looked like the vanguard of a particularly spiffy occupying force.

During my first year as provost at Buffalo, I recruited nine new chairs and two new deans, changing about 90 percent of the leadership structure of the social sciences. The faculty gained forty-five new full-time faculty (almost 75 percent of the present Buffalo faculty were appointed under Meyerson). I personally interviewed more than thirty candidates for various jobs. The newcomers were a largely self-selected group who shared the commitment to innovation, risk taking, and excellence that was the credo of the Meyerson presidency.

For that first year Buffalo was a kind of academic Camelot. When we provosts gathered around the president's conference table, we were ready to work miracles. Occasionally, however, signals reached me that not everyone took us quite as seriously as we took ourselves. One morning I found that on my coatrack someone had hung a Batman cape (I eschewed the down parka that was the winter uniform of the campus and wore a Tyrolean cape that I thought was quite dashing). The anonymous critic had a point. Omnipotent fantasy was the delusion of choice that year in the administration building.

The occasional doubt crept in. In my end-of-the-year provost's report, I ticked off the high points of the ambitious reorganization we were undertaking but cautioned that "each of these virtues could be transformed overnight into obstacles and problems." But such reservations were rare.

The commitment to transforming the university shaped my home life as well. Our house was constantly filled with academic superstars and promising newcomers we were desperately trying to recruit and with Buffalonians with whom we were frantically getting acquainted. As one local wag observed, my wife and I were always entertaining "two hundred of [our] closest friends." That first year we had sixty-five parties—brunches in the garden, afternoon wine-and-cheese parties, but mostly large dinner parties where everyone who mattered was invited to meet whoever was being wooed.

Sometime late in that first spring, my four-year-old daughter, Kate, came into the garden while I was reading the newspapers in a rare moment of stillness. She looked pensive. I asked her what she was thinking. She said she was thinking about what she wanted to do when she grew up.

"What have you decided?" I asked.

She paused for a long time and then said, "When I grow up, I want to be a guest."

I picked her up and laughed. And then later, when I was by myself, I cried. At four Kate was like the psychiatrist's child who wants to grow up to be a patient. Her remark wasn't so much clever as it was true—and terribly sad.

I learned a great deal at Buffalo, but one thing I did not learn was how to integrate intimacy with ambition. I still haven't learned at sixty-eight. Yeats writes that "the intellect of man is forced to choose / Perfection of the life, or of the work, and / if it take the second must refuse / A heavenly mansion, raging in the dark." In real life the dilemma is even more excruciating, because it is often the people we love who are left raging in the dark.

Like John Kennedy's Camelot, our academic utopia lasted roughly a thousand days. By 1970 our attempt to transform the university was interrupted by the campus unrest that was sweeping the country. At one point six hundred police officers in full riot gear appeared on campus, ready to use force if protesting students got too uppity. It was not academic life as any of us had known it. A student filmmaker at Buffalo put it nicely when he titled his cinema vérité record of that turbulent semester *Andy Hardly Goes to College*. The halls of ivy were no longer filled with the optimism of a few years earlier but instead were filled with the lingering smell of tear gas.

In the final analysis the Meyersonian spirit at Buffalo was defeated by a changed political and economic reality. Yet there were ways in which we contributed, however unwittingly, to that failure.

Examining what went wrong at Buffalo altered forever the way I think about change. Martin Meyerson had the first thing that every effective leader needs—a powerful vision of the way the organization should be, a vision he was able to communicate to me and many of his other recruits. But unless a vision is sustained by action, it quickly turns to ashes.

The Meyersonian dream never got out of the administration building. In ways that only later became clear, we undermined the very thing we wanted most. Our actions and even our style tended to alienate the people who would be most affected by the changes we proposed. Failing to appreciate the importance to the organization of the people who are already in it is a classic managerial mistake, one that new managers and change-oriented administrators are especially prone to make. We certainly did. In our Porsches and berets, we acted as if the organization hadn't existed until the day we arrived.

There are no clean slates in established organizations. A new administration can't play Noah and build the world anew with two handpicked representatives of each academic discipline. Talk of new beginnings is so much rhetoric—frightening rhetoric to those who suspect that the new signals the end of their own careers. At Buffalo we newcomers disregarded history. But without history, without continuity, there can be no successful change. A.N. Whitehead said it best: "Every leader, to be effective, must simultaneously adhere to the symbols of change and revision and the symbols of tradition and stability."

What most of us in organizations really want (and what status, money, and power serve as currency for) is acceptance, affection, self-esteem. Institutions are more amenable to change when the esteem of all members is preserved and enhanced. Whatever people say, given economic sufficiency they stay in organizations and feel satisfied in them because they feel competent and valued. Change carries the threat of loss. When managers remove that threat, people are much freer to identify with the adaptive process and much better equipped to tolerate the high degree of ambiguity that accompanies change.

When I think of Buffalo, I think of that joke "How many psychiatrists does it take to change a light bulb?" The answer is "One, but the light bulb really has to want to change." Organizations change themselves when the members want to. You can't force them to change, even in a Batman cape.

The University of Cincinnati: 1971–1978

The logical next step after Buffalo was a college presidency. My name began to surface at the meetings of presidential search committees, short-lived organizations I would come to know well over the next two decades. During 1970–1971 my name appeared on several short lists, and in the fall of 1971 I became president of the University of Cincinnati.

Less than a year into my tenure, I had a moment of truth. I was sitting in my office on campus, mired in the incredible stack of paperwork on my desk. It was four o'clock in the morning. Weary of bone and tired of soul, I found myself muttering, "Either I can't manage this place, or it's unmanageable."

As I sat there, I thought of a friend and former colleague who had become president of one of the nation's top universities. He had started out full of fire and vision. But a few years later, he had quit. "I never got around to doing the things I wanted to do," he explained.

Sitting there in the echoing silence, I realized that I had become the victim of a vast, amorphous, unwitting conspiracy to prevent me from doing anything whatsoever to change the status quo. Unfortunately, I was one of the chief conspirators. This discovery caused me to formulate what I thought of as Bennis's First Law of Academic Pseudodynamics, which states that routine work drives out nonroutine work and smothers to death all creative planning, all fundamental change in the university—or any institution, for that matter.

The evidence surrounded me. To start, there were 150 letters in the day's mail that required a response. About a third of them concerned our young dean of education, Hendrik Gideonse. His job was a critical one—to bring about change in the way the university taught teachers and to create a new relationship between the university and the precollege students in the deprived and deteriorating neighborhood around us, the neighborhood from which we drew an increasing number of students.

But the letters were not about education. They were about a baby, Gideonse's ten-week-old son. The young dean was committed to ensuring that his wife had the time and freedom to develop her potential as fully as his own. And so he was carrying the baby to his office twice a week in a portable bassinet that he kept on the desk while he worked. The local paper had run a story on Gideonse and his young office companion, with picture, on page one. National TV had picked up the story. As a result, my in-basket had begun to overflow with letters urging me to dismiss the dean or at least have him arrested for child abuse. My response was to say that if Gideonse could engage in that form of applied humanism and still accomplish the things we both wanted in education, then I, like Lincoln with Grant's whiskey, would gladly send him several additional babies for adoption. But there was no question that Hendrik and his baby took up quite a bit of my time.

Also on my desk was a note from a professor, complaining that his classroom temperature was down to sixty-five degrees. Someone once observed that trying to lead faculty is like herding cats. What did this man expect me to do—grab a wrench and fix the heating system myself? A parent complained about a four-letter word in a Philip Roth book being used in an English class. The track coach wanted me to come over and see for myself how bad the track was.

And that was the easy stuff. That year perhaps 20 percent of my time had been taken up by a problem at the general hospital, which was owned by the city but administered by the university and which served as the teaching hospital of the UC medical school. A group of terminal-cancer patients had, with their consent, been subjected to whole-body radiation as a possible beneficial therapy. The Pentagon, interested in gauging the human effects of nuclear warfare, had helped subsidize the study.

Like Hendrik's baby, this too became a major story, one in

which irresponsible comparisons were made between the cancer study and Nazi experiments on human guinea pigs. The flap eventually subsided after a blue-ribbon task force recommended changes in the experiment's design. But by then I had invested endless time in a matter only vaguely related to the primary purposes of the university—and wound up being accused of interfering with academic freedom in the bargain.

In my cluttered office that morning, I grew up in some fundamental way. I realized that, from now on, my principal role model was going to have to be me. I decided that the kind of university president I wanted to be was one who led, not managed. That's an important difference. Many an institution is well managed yet very poorly led. It excels in the ability to handle all the daily routine inputs yet never asks whether the routine should be done in the first place.

My entrapment in minutiae made me realize another thing: that people were following the old army game. They did not want to take responsibility for the decisions they properly should make. "Let's push up the tough ones" had become the motto. As a result, everybody was dumping his or her "wet babies" (as old hands at the State Department call them) on my desk. I decided then and there that my highest priority was to create an "executive constellation" to run the office of the president. The sole requirements for inclusion in the group were that the individual needed to know more than I did about his or her area of competence and had to be willing to take care of daily matters without referring them back to me. I was going to make the time to lead.

I realized that I had been doing what so many leaders do: I was trying to be everything to the organization—father, fixer, policeman, ombudsman, rabbi, therapist, and banker. As a CEO who was similarly afflicted put it to me later, "If I'm walking on the shop floor and see a leak in the dike, I have to stick my finger in." Trying to be everything to everyone was diverting me from real leadership. It was burning me out. And perhaps worst of all, it was denying all the potential leaders under me the chance to learn and prove themselves.

Things got better after that, although I never came close to the ideal. As I look back at my experience at UC, I compare it with my psychoanalysis: I wouldn't have missed it for the world, and I would never go through it again. In becoming a leader I learned a number of important things about both leadership and myself. As Sophocles observes in *Antigone*, "But hard it is to learn the mind of any mortal, or the heart, 'til he be tried in chief authority. Power shows the man."

Having executive power showed me three personal truths.

First, I was, as the song says, "looking for love in all the wrong places." Intellectually I knew that leaders can't, shouldn't, count on being loved. But I seriously underestimated the emotional impact of angry constituents. I believed the false dream that people would love me if only they really got to know me. I call it the Lennie Bernstein syndrome. Ned Rorem, Bernstein's friend and colleague, recalls how "Lennie" was furious about a negative review in the *New York Times*. "He hates me," Bernstein said of the critic. Rorem suggested gently that Bernstein really couldn't expect everyone to love him. Bernstein was stunned for a moment by his friend's insight. "Oh, yeah," Bernstein finally conceded, "that's because you can't meet everybody."

Even worse than not being loved was not being understood. I found that so dispiriting that I began to develop a whole new theory about the social determinants of depression. It came to me sometime in 1976. I was flying back from a fund-raising trip to Washington, D.C., and idly skimming through the pages of the American Airlines magazine *American Way,* when I came across a fascinating feature about the items various famous Americans would leave in a time capsule to symbolize America on its two hundredth anniversary. The first celebrity was the astronaut Neil Armstrong. I had helped recruit Armstrong to UC and was floored by his response. He chose a credit card (I forget what it was supposed to symbolize).

I was simply amazed that Armstrong thought the credit card was emblematic of the United States in 1976. It seemed so remote from the technological triumph the country had accomplished in space or what Armstrong himself was famous for. By sheer happenstance, when I got back to Cincinnati that night I attended a party to which Armstrong had also been invited. I couldn't resist asking him about the magazine article. I told him I couldn't understand why he hadn't brought up his famous moonwalk or something else about his historic space flight. Why hadn't he chosen the American flag he planted on the moon or his fabled camera instead of a credit card?

He looked at me sadly and answered as if from the Slough of Despond. "You too," he said. "Isn't there any way I can escape the astronaut image? Do you realize that I've been teaching aeronautical engineering at your university for the past five years, getting decent student ratings, working on a few important bioengineering projects at the medical school, and still, all you think of me as is Neil Armstrong, astronaut? No wonder I'm depressed."

Anyone in authority, astronaut or baseball player, university president or national leader, is to some extent the hostage of how others perceive him or her. The perceptions of other people can be a prison. For the first time I began to understand what it must be like to be the victim of prejudice, to be helpless in the steel embrace of how other people see you. People impute motives to their leaders, love or hate them, seek them out or avoid them, and idolize or demonize them independently of what the leaders do or are. Ironically, at the very time I had the most power, I felt the greatest sense of powerlessness.

Finally, at UC I began to work out the relationship every leader must resolve between the self and the organization. One university president I knew took his own life because, according to his best friend, "he cared too much for the institution." I recall giving a lecture at Harvard and going on and on about the difficulties of presiding over a university. Afterward Paul Ylvisaker, then the dean of Harvard's school of education, asked me, "Warren, do you love the university enough?" It was an unsettling question. I said I didn't know for sure and would have to think a lot more about it.

Secretly I had doubts about how much I loved UC. I felt that my predecessor had cared too much for the university, so much so that he thought, in the manner of de Gaulle, "UC, *c'est moi*." I think it's dangerous to identify so strongly with an institution that your own self-esteem can be affected by the outcome of the campus Big Game.

Ultimately, I think, a leader should love the organization enough to help create a self-activating life for it. He or she has to love it enough to try to turn it into an environment in which others can understand and care for it, even in difficult times. A leader has to care enough about the organization to want it to be autonomous, able to function very nicely without him or her.

And I realized an important personal truth. I was never going to be completely happy with positional power, the only kind of power an organization can bestow. What I really wanted was personal power, influence based on voice.

The Rest So Far

In February 1979 I joined forty other scholars, executives, and futurologists at Windsor Castle for a colloquium on the evolutionary forces at work in society and the ways in which rapid change was likely to affect management.

I learned more about change that week than I had bargained for. At the age of fifty-three I experienced a myocardial infarction, a heart attack that landed me in a fifteen-bed ward at historic Middlesex Hospital, with a glamorous society photographer on my right and an engaging tramp on my left. It was probably the most crucial event of my adult life. I eventually spent three months in England, recuperating. And during that period I had nothing to do but think about what I had learned in the course of five decades and what I wanted to do with the rest of my life.

During the avoidance phase of my recovery, while I was groping for distractions, I learned that Rudyard Kipling had been taken to Middlesex Hospital with a perforated duodenum. When his physician asked Kipling what was the matter, he explained, "Something has come adrift inside" (he died in Middlesex a week later).

I too was acutely aware that something had come adrift inside. Forcibly removed from the overbooked professional life I had created for myself, I began to write poetry for the first time ever.

Once again I was discovering what I had learned by putting it into words. My favorite of the poems was one called "Plea Bargaining." In it an authoritative voice asks, "How soon would you like to visit your grave? A. Not for a long, long time."

During my recuperation at Windsor Castle, I received a call from Jim O'Toole, a management professor at USC, wondering if I would be interested in a professorship there. In June 1979 I visited the campus. I was intrigued by then-dean Jack Steele's vision for the School of Business Administration (I'm a sucker for a vision every time) and astonished by the quality of the faculty he had assembled.

I am now in my eleventh year at USC—my longest continuous tenure at any institution. In many ways it has been the happiest period of my life. USC has provided me with exactly the right social architecture to do what seems most important to me now: teaching in the broadest sense.

At USC I have the leisure to consolidate what I've learned—about self-invention, about the importance of organization, about the nature of change, about the nature of leadership—and to find ways to communicate those lessons. Erik Erikson talks about an eight-stage process of human development. I think I have entered Erikson's seventh stage—the generative one—in which self-absorption gives way to an altruistic surrender to the next generation. Although writing is my greatest joy, I also take enormous pleasure in people-growing, in watching others bloom, in mentoring as I was mentored.

USC has provided me with several other structures for transmitting what I know, including a new Leadership Institute that will be an international center for the study of leadership and the development of leaders in every field. Recently I was also able to apply the sometimes painful lessons I learned as a university president, in the course of heading the committee that chose Steven Sample to be the new USC president.

My father died when he was fifty-nine, as did Doug McGregor. Abe Maslow was barely sixty. When I was growing up, that was the norm. But now, at sixty-eight, I find that I need a new role model for the last third of my life, which is shaping up to be the most challenging of all. In a recent television commentary, writer John Leonard praised his mother and his mother-in-law, two remarkable women in their eighties. He said of them that in the course of their lives, they had been pushed out of the windows of a lot of burning buildings. "I need to know," said Leonard, "how they learned to bounce."

I am learning how to bounce from my closest friend, Sam Jaffe, who will be ninety-two this year. (When in his fifties Sam was the Academy Award–winning producer of *Born Free*.) What I have already discovered is that the need to reinvent oneself, to "compose a life," as Mary Catherine Bateson puts it, is ongoing. Three years ago Sam and I took a summer course on Dickens at Trinity Hall, Cambridge. Sam, who recently tried to buy the film rights to a book I had given him, continues to scrimmage in the notoriously competitive subculture of Hollywood. He gives me hope.

I find that I have acquired a new set of priorities. Some of the old agonies have simply disappeared. I have no doubt that my three children are more important than anything else in my life. Having achieved a certain level of worldly success, I need hardly think about it anymore. Gentler virtues seem terribly important now. I strive to be generous and productive. I would hope to be thought of as a decent and creative man.

I think Miss Shirer would be proud.

2

Is Democracy Inevitable?

Twenty-six years after this essay was originally published, it suddenly came true. The preface written for its republication in 1990 in the Harvard Business Review *places the piece in context and is also included. When Phil Slater and I first hypothesized that democracy would eventually triumph because it worked, we didn't instantly convert the world, or even our editor. Our proposed title, "Democracy Is Inevitable," was changed to a more cautious "Is Democracy Inevitable?" When I saw, via CNN, the Berlin Wall begin to crumble, it was all I could do to keep from shouting, "Yes!"*

Phil Slater was on the Harvard faculty when we wrote this essay. He is now an author and artistic director of the Santa Cruz County Actors' Theatre.

Retrospective Commentary

It's wonderful—perhaps because it's so rare—to reread something you wrote 26 years ago and discover you were right.

In 1990, after the extraordinary recent events in Eastern Europe, including the dismantling of the Berlin Wall, it seems obvious that democracy was inevitable. But 26 years ago, in the heat of the Cold War, it was not so certain. When Philip Slater and I first argued that democracy would eventually dominate in both the world and the workplace, a nuclear war between the United States and the Soviet Union seemed more likely than a McDonald's in Moscow.

Slater and I saw a common thread running through the most exciting organizations of the time: as the once-absolute power of top management atrophied, a more collegial organization where good ideas were valued even if they weren't the boss's was emerging. We were convinced democracy would triumph for a simple but utterly compelling reason—it worked. It was, and is, more effective than autocracy, bureaucracy, and other nondemocratic forms of organization.

It is only fair to note that in international politics, democratization is a very recent phenomenon, although a profound one. Only a year ago, Nicolae Ceausescu had the power to ban birth control in Romania and require that every typewriter be registered. The state even regulated the temperature of Romanian households. The events of recent months are even more remarkable because they were so long in coming. It was easier to speculate 26 years ago that democracy was inevitable than to imagine five months ago that the notoriously repressive military government of Myanmar, formerly Burma, would be ousted peacefully by the National League for Democracy, as it was in May of this year.

The democratization of the workplace has made fewer headlines but has been no less dramatic. In the 1960s, participative management was a radical enough notion that some of the Sloan fellows at MIT accused me of being a Communist for espousing it. Now most major corporations practice some form of egalitarian management. The pyramid-shaped organization chart has gone the way of the Edsel.

The change is pervasive. Self-managed work groups are replacing assembly lines in auto plants. Organizations as disparate as Herman Miller, the manufacturer of office furniture, and Beth Israel Hospital in Boston have adopted the democratic management techniques of the late Joseph Scanlon, one of the first to appreciate that employee involvement is crucial for quality control. At Hewlett-Packard's facility in Greeley, Colorado, most decisions are made not by traditional managers but by frontline employees who work in teams on parts of projects. Even project coordination is done by team representatives, working on committees known as "boards of directors."

No longer a monolith, the successful modern corporation is a Lego set whose parts can be readily reconfigured as circumstances change. The old paradigm that exalted control, order, and predictability has given way to a nonhierarchical order in which all employees' contributions are solicited and acknowledged and in which creativity is valued over blind loyalty. Sheer self-interest motivated the change. Organizations that encourage broad participation, even dissent, make better decisions. In a recent study, Rebecca A. Henry, a psychology professor at Purdue University, found that groups are better forecasters than are individuals. And the more the group disagrees initially, the more accurate the forecast is likely to be.

Slater and I were right on target, I think, in writing both that adaptability would become the most important determinant of an organization's survival and that information would drive the organization of the future. The person who has information wields more power than ever before. And even though industrial applications of the computer were still in their first decade, we sensed how important processing technology would be, largely, I suspect, because we were working in the Boston area, the birthplace of so much of the new technology.

I don't think we fully appreciated, however, the extent to which the new technology would accelerate the pace of change and help create a global corporation, if not a global village. With computers and fax machines, New York Life Insurance processes its claims not in New York or even the United States but in Ireland. Several years ago, I invited the Dalai Lama to participate in a seminar for CEOs at the University of Southern California. The embodiment of thousands of years of Tibetan spiritualism graciously declined by fax.

Slater and I failed to foresee one development that would profoundly change organizational life: the extraordinary role Japan would play in shaping U.S. corporate behavior in the 1980s. The discovery that another nation could challenge U.S. dominance in the marketplace inspired massive self-evaluation and forever disrupted the status quo. Nothing contributed more to the democratization of business than the belief, true or false, that Japanese management was more consensual than U.S. management. To meet Japanese competition, U.S. leaders were willing to do anything, even share their traditional prerogatives with subordinates.

So a new kind of leader has emerged who is a facilitator, not an autocrat, an appreciator of ideas, not necessarily a fount of them. The Great Man—or Woman—still exists as the public face of companies and countries, but the leader and the organization are no longer one and the same.

Around the world, the generals are being ousted and the poets are taking charge. Slater and I argued that the military-bureaucratic model was increasingly obsolete and was being replaced by a scientific model. That is still true. Science not only tolerates change; it creates change. And, as we wrote, science flourishes only in a democracy, the one form of organization recognizing that creativity, an invaluable commodity, is utterly unpredictable and can come from any quarter.

Warren G. Bennis
(1990)

Cynical observers have always been fond of pointing out that business leaders who extol the virtues of democracy on ceremonial occasions would be the last to think of applying them to their own organizations. To the extent that this is true, however, it reflects a state of mind that is by no means peculiar to businesspeople but characterizes all Americans, if not perhaps all citizens of democracies.

This attitude is that democracy is a nice way of life for nice people, despite its manifold inconveniences—a kind of expensive and inefficient luxury, like owning a large medieval castle. Feelings about it are for the most part affectionate, even respectful, yet a little impatient. There are probably few people in the United States who have not at some time nourished in their hearts the blasphemous thought that life would go much more smoothly if democracy could be relegated to some kind of Sunday morning devotion.

The bluff practicality of the "nice but inefficient" stereotype masks a hidden idealism, however, for it implies that institutions can survive in a competitive environment through the sheer good-heartedness of those who maintain them. We challenge this notion. Even if all those benign sentiments were eradicated today, we would awaken tomorrow to find democracy still entrenched, buttressed by a set of economic, social, and political forces as practical as they are uncontrollable.

Democracy has been so widely embraced not because of some vague yearning for human rights but because *under certain conditions* it is a more "efficient" form of social organization. (Our concept of efficiency includes the ability to survive and prosper.) It is not accidental that those nations of the world that have endured longest under conditions of relative wealth and stability are democratic, while authoritarian regimes have, with few exceptions, either crumbled or eked out a precarious and backward existence.

Despite this evidence, even so acute a statesman as Adlai Stevenson argued in a *New York Times* article on November 4, 1962, that the goals of the Communists are different from ours. "They are interested in power," he said, "we in community. With such fundamentally different aims, how is it possible to compare communism and democracy in terms of efficiency?"

Democracy (whether capitalistic or socialistic is not at issue here) is the only system that can successfully cope with the changing demands of contemporary civilization. We are not necessarily endorsing democracy as such; one might reasonably argue that industrial civilization is pernicious and should be abolished. We suggest merely that given a desire to survive in this civilization, democracy is the most effective means to this end.

Democracy Takes Over

There are signs that our business community is becoming aware of democracy's efficiency. Several of the newest and most rapidly blooming companies in the United States boast unusually democratic organizations. Even more surprising, some of the largest established corporations have been moving steadily, if accidentally, toward democratization. Feeling that administrative vitality and creativity were lacking in their systems of organization, they enlisted the support of social scientists and outside programs. The net effect has been to democratize their organizations. Executives and even entire management staffs have been sent to participate in human relations and organizational laboratories to learn skills and attitudes that ten years ago would have been denounced as anarchic and revolutionary. At these meetings, status prerogatives and traditional concepts of authority are severely challenged.

Many social scientists have played an important role in this development. The contemporary theories of McGregor, Likert, Argyris, and Blake have paved the way to a new social architecture. Research and training centers at the National Training Laboratories, Tavistock Institute, Massachusetts Institute of Technology, Harvard Business School, Boston University, University of California at Los Angeles, Case Institute of Technology, and others have pioneered in applying social science knowledge to improving organizational effectiveness. The forecast seems to hold genuine promise of progress.

System of Values

What we have in mind when we use the term "democracy" is not "permissiveness" or "laissez-faire" but a system of values—a climate of beliefs governing behavior—that people are internally compelled to affirm by deeds as well as words. These values include:

1. Full and free *communication,* regardless of rank and power.

2. A reliance on *consensus* rather than on coercion or compromise to manage conflict.

3. The idea that *influence* is based on technical competence and knowledge rather than on the vagaries of personal whims or prerogatives of power.

4. An atmosphere that permits and even encourages emotional *expression* as well as task-oriented behavior.

5. A basically *human* bias, one that accepts the inevitability of conflict between the organization and the individual but is willing to cope with and mediate this conflict on rational grounds.

Changes along these dimensions are being promoted widely in U.S. industry. Most important for our analysis is what we believe to be the reason for these changes: *democracy becomes a functional necessity whenever a social system is competing for survival under conditions of chronic change.*

Adaptability to Change

Technological innovation is the most familiar variety of such change to the inhabitants of the modern world. But if change has now become a permanent and accelerating factor in American life, then adaptability to change becomes the most important determinant of survival. The profit, the saving, the efficiency, and the morale of the moment become secondary to keeping the door open for rapid readjustment to changing conditions.

Organization and communication research at MIT reveals quite dramatically what type of organization is best suited for which kind of environment. Specifically:

• For simple tasks under static conditions, an autocratic, centralized structure, such as has characterized most industrial organizations in the past, is quicker, neater, and more efficient.

• But for adaptability to changing conditions, for "rapid acceptance of a new idea," for "flexibility in dealing with novel problems, generally high morale and loyalty . . . the more egalitarian or decentralized type seems to work better." One of the reasons for this is that the centralized decision maker is "apt to discard an idea on the grounds that he is too busy or the idea too impractical."[1]

Our argument for democracy rests on an additional factor, one that is fairly complicated but profoundly important. Modern industrial organization has been based roughly on the antiquated system of the military. Relics of this can still be found in clumsy terminology such as "line and staff," "standard operating procedure," "table of organization," and so on. Other remnants can be seen in the emotional and mental assumptions regarding work and motivation held today by some managers and industrial consultants. By and large, these conceptions are changing, and even the military is moving away from the oversimplified and questionable assumptions on which its organization was originally based. Even more striking, as we have mentioned, are developments taking place in industry, no less profound than a fundamental move from the autocratic and arbitrary vagaries of the past toward democratic decision making.

This change has been coming about because of the palpable inadequacy of the military-bureaucratic model, particularly its response to rapid change, and because the institution of science is now emerging as a more suitable model.

Why is science gaining acceptance as a model? Not because we teach and conduct research within research-oriented universities. Curiously enough, universities have been resistant to democratization, far more so than most other institutions.

Science is winning out because the challenges facing modern enterprises are *knowledge*-gathering, *truth*-requiring dilemmas. Managers are not scientists, nor do we expect them to be. But the processes of problem solving, conflict resolution, and recognition of dilemmas have great kinship with the academic pursuit of truth. The institution of science is the only institution based on and geared for change. It is built not only to adapt to change but also to overthrow and create change. So it is—and will be—with modern industrial enterprises.

And here we come to the point. For the spirit of inquiry, the foundation of science, to grow and flourish, there must be a democratic environment. Science encourages a political view that is egalitarian, pluralistic, liberal. It accentuates freedom of opinion and dissent. It is against all forms of totalitarianism, dogma, mechanization, and blind obedience. As a prominent social psychologist has pointed out, "Men have asked for freedom, justice, and respect precisely as science has spread among them."[2] In short, the only way organizations can ensure a scientific attitude is to provide the democratic social conditions where one can flourish.

In other words, democracy in industry is not an idealistic conception but a hard necessity in those areas where change is ever present and creative scientific enterprise must be nourished. For democracy is the only system of organization that is compatible with perpetual change.

Retarding Factors

It might be objected here that we have been living in an era of rapid technological change for a hundred years, without any noticeable change in the average industrial company. True, there are many restrictions on the power of executives over their subordinates today compared with those prevailing at the end of the nineteenth century. But this hardly constitutes industrial democracy—the decision-making function is still an exclusive and jealously guarded prerogative of the top echelons. If democracy is an inevitable consequence of perpetual change, why have we not seen more dramatic changes in the structure of industrial organizations? The answer is twofold.

Obsolete Individuals

First, technological change is rapidly accelerating. We are now beginning an era when people's knowledge and approach can become obsolete before they have even begun the careers for which they were trained. We are living in an era of runaway inflation of knowledge and skill, where the value of what one learns is always slipping away. Perhaps this explains the feelings of futility, alienation, and lack of individual worth that are said to characterize our time.

Under such conditions, the individual *is* of relatively little significance. No matter how imaginative, energetic, and brilliant individuals may be, time will soon catch up with them to the point where they can profitably be replaced by others equally imaginative, energetic, and brilliant but with a more up-to-date viewpoint and fewer obsolete preconceptions. As Martin Gardner says about the difficulty some physicists have in grasping Einstein's theory of relativity: "If you are young, you have a great advantage over these scientists. Your mind has not yet developed those deep furrows along which thoughts so often are forced to travel."[3] This situation is just beginning to be felt as an immediate reality in U.S. industry, and it is this kind of uncontrollably rapid change that generates democratization.

Powers of Resistance

The second reason is that the mere existence of a dysfunctional tendency, such as the relatively slow adaptability of authoritarian structures, does not automatically bring about its disappearance. This drawback must first either be recognized for what it is or become so severe as to destroy the structures in which it is embedded. Both conditions are only now beginning to make themselves felt, primarily through the peculiar nature of modern technological competition.

The crucial change has been that the threat of technological defeat no longer comes necessarily from rivals within the industry, who usually can be imitated quickly without too great a loss, but often comes from outside—from new industries using new materials in new ways. One can therefore make no intelligent prediction about the next likely developments in industry. The blow may come from anywhere. Correspondingly, a viable corporation cannot merely develop and advance in the usual ways. To survive and grow, it must be prepared to go anywhere—to develop new products or techniques even if they are irrelevant to the present activities of the organization.[4] Perhaps that is why the beginnings of democratization have appeared most often in industries that depend heavily on invention, such as electronics. It is undoubtedly why more and more sprawling behemoths are planning consequential changes in their organizational structures and climates to release democratic potential.

Farewell to "Great Men"

The passing of years has also given the coup de grace to another force that retarded democratization—the "great man" who with brilliance and farsightedness could preside with dictatorial powers at the head of a growing organization and keep it at the vanguard of U.S. business. In the past, this person was usually a man with a single idea, or a constellation of related ideas, which he developed brilliantly. This is no longer enough.

Today, just as he begins to reap the harvest of his imagination, he finds himself suddenly outmoded because someone else (even perhaps one of his stodgier competitors, aroused by desperation) has carried the innovation a step further or found an entirely new and superior approach to it. How easily can he abandon his idea, which contains all his hopes, his ambitions, his very heart? His aggressiveness now begins to turn in on his own organization; and the absolutism of his position begins to be a liability, a dead hand on the flexibility and growth of the company. But the great man cannot be removed. In the short run, the company would even be hurt by his loss since its prestige derives to such an extent from his reputation. And by the time he has left, the organization will have receded into a secondary position within the industry. It might decay further when his personal touch is lost.

The "cult of personality" still exists, of course, but it is rapidly fading. More and more large corporations (General Motors, for one) predicate their growth not on "heroes" but on solid management teams.

Organization Men

Taking the place of the "great man," we are told, is the "organization man." Liberals and conservatives alike have shed many tears over this transition. The liberals have in mind as "the individual" some sort of creative deviant—an intellectual, artist, or radical politician. The conservatives are thinking of the old captains of industry and perhaps some great generals.

Neither is at all unhappy to lose the "individuals" mourned by the other, dismissing them contemptuously as Communists and rabblerousers on the one hand, criminals and Fascists on the other. What is particularly confusing in terms of the present issue is a tendency to equate conformity with autocracy—to see the new industrial organization as one in which all individualism is lost except for a few villainous, individualistic manipulators at the top.

But this, of course, is absurd in the long run. The trend toward the "organization man" is also a trend toward a looser and more flexible organization in which the roles to some extent are interchangeable and no one is indispensable. To many people, this trend is a monstrous nightmare, but one should not confuse it with the nightmares of the past. It may mean anonymity and homogeneity, but it does not and cannot mean authoritarianism, despite the bizarre anomalies and hybrids that may arise in a period of transition.

The reason it cannot is that it arises out of a need for flexibility and adaptability. Democracy and the dubious trend toward the "organization man" alike (for this trend is a part of democratization, whether we like it or not) arise from the need to maximize the availability of appropriate knowledge, skill, and insight under conditions of great variability.

Rise of the Professional

While the "organization man" idea has titillated the public imagination, it has masked a far more fundamental change now taking place: the rise of the "professional." Professional specialists, holding advanced degrees in such abstruse sciences as cryogenics or computer logic as well as in the more mundane business disciplines, are entering all types of organizations at a higher rate than any other sector of the labor market.

Such people seemingly derive their rewards from inner standards of excellence, from their professional societies, from the intrinsic satisfaction of their tasks. In fact, they are committed to the task, not the job; to their standards, not their boss. They are uncommitted except to the challenging environments where they can "play with problems."

These new professionals are remarkably compatible with our conception of a democratic system. For like them, democracy seeks no new stability, no end point; it is purposeless, save that it purports to ensure perpetual transition, constant alteration, ceaseless instability. It attempts to upset nothing, but only to facilitate the potential upset of anything. Democracy and professionals identify primarily with the adaptive process, not the "establishment."

Yet all democratic systems are not entirely so—there are always limits to the degree of fluidity that can be borne. Thus it is not a contradiction to the theory of democracy to find that a particular democratic society or organization may be more "conservative" than an autocratic one. Indeed, the most dramatic, violent, and drastic changes have always taken place under autocratic regimes, for such changes usually require prolonged self-denial, while democracy rarely lends itself to such voluntary asceticism. But these changes have been viewed as finite and temporary, aimed at a specific set of reforms and moving toward a new state of nonchange. It is only when the society reaches a level of technological development at which survival is dependent on the institutionalization of perpetual change that democracy becomes necessary.

Reinforcing Factors

The Soviet Union is rapidly approaching this level and is beginning to show the effects, as we shall see. The United States has already reached it. Yet democratic institutions existed in the United States when it was still an agrarian nation. Indeed, democracy has existed in many places and at many times, long before the advent of modern technology. How can we account for these facts?

Expanding Conditions

First, it must be remembered that modern technology is not the only factor that could give rise to conditions of necessary perpetual change. Any situation involving rapid and unplanned expansion sustained over a sufficient period of time will tend to produce great pressure for democratization. Secondly, when we speak of democracy, we are referring not exclusively or even primarily to a particular political format. Indeed, American egalitarianism has perhaps its most important manifestation not in the Constitution but in the family.

Historians are fond of pointing out that Americans have always lived under expanding conditions—first the frontier, then the successive waves of immigration, now a runaway technology. The social effects of these kinds of expansion are of course profoundly different in many ways, but they share one impact: all have made it impossible for an authoritarian family system to develop on a large scale. Every foreign observer of American mores since the seventeenth century has commented that American children "have no respect for their parents," and every generation of Americans since 1650 has produced forgetful native moralists complaining about the decline in filial obedience and deference.

Descriptions of family life in colonial times make it quite clear that American parents were as easygoing, permissive, and child oriented then as now, and the children as independent and disrespectful. This lack of respect is not for the parents as individuals but for the concept of parental authority as such.

The basis for this loss of respect has been outlined quite dramatically by historian Oscar Handlin, who points out that in each generation of early settlers, the children were more at home in their new environment than their parents were—had less fear of the wilderness, fewer inhibiting European preconceptions and habits.[5] Furthermore, their parents were heavily dependent on them physically and economically. This was less true of the older families after the East became settled. But nearer the frontier, the conditions for familial democracy became again strikingly marked so that the cultural norm was protected from serious decay.

Further reinforcement came from new immigrants, who found their children better adapted to the world because of their better command of the language, better knowledge of the culture, better occupational opportunities, and so forth. It was the children who were expected to improve the social position of the family and who through their exposure to peer groups and the school system could act as intermediaries between their parents and the new world. It was not so much "American ways" that shook up the old family patterns as the demands and requirements of a new situation. How could the young look to the old as the ultimate fount of wisdom and knowledge when, in fact, their knowledge was irrelevant—when the children indeed had a better practical grasp of the realities of American life than did their elders?

The New Generation

These sources of reinforcement have now disappeared. But a third source has only just begun. Rapid technological change again means that the wisdom of elders is largely obsolete and that the young are better adapted to their culture than are their parents.

This fact reveals the basis for the association between democracy and change. The old, the learned, the powerful, the wealthy, those in authority—these are the ones who are committed. They have learned a pattern and succeeded in it. But when change comes, it is often the *uncommitted* who can best realize it and take advantage of it. This is why primogeniture has always lent itself so easily to social change in general and to industrialization in particular. The uncommitted younger children, barred from success in the older system, are always ready to exploit new opportunities. In Japan, younger sons were treated more indulgently by their parents and given more freedom to choose an occupation since "in Japanese folk wisdom, it is the younger sons who are the innovators."[6]

Democracy is a superior technique for making the uncommitted more available. The price it extracts is uninvolvement, alienation, and skepticism. The benefits that it gives are flexibility and the joy of confronting new dilemmas.

Doubt and Fears

Indeed, we may even in this way account for the poor opinion democracy has of itself. We underrate the strength of democracy because it creates a general attitude of doubt, skepticism, and modesty. It is only among the authoritarian that we find the dogmatic confidence, the self-righteousness, the intolerance and cruelty that permit one never to doubt oneself and one's beliefs. The looseness, sloppiness, and untidiness of democratic structures express the feeling that what has been arrived at today is probably only a partial solution and may well have to be changed tomorrow.

In other words, one cannot believe that change is in itself a good thing and still believe implicitly in the rightness of the present. Judging from the report of history, democracy has always underrated itself—one cannot find a democracy anywhere without also discovering (side by side with expressions of outrageous chauvinism) an endless pile of contemptuous and exasperated denunciations of it. (One of the key issues in our national politics today, as in the presidential campaign in 1960, is our "national prestige.") And perhaps this is only appropriate. For when a democracy ceases finding fault with itself, it has probably ceased to be a democracy.

Overestimating Autocracy

But feeling doubt about our own social system need not lead us to overestimate the virtues and efficiency of others. We can find this kind of overestimation in the exaggerated fear of the "Red Menace"—mere exposure to which is seen as leading to automatic conversion. Few authoritarians can conceive of the possibility that an individual could encounter an authoritarian ideology and not be swept away by it.

More widespread is the "better dead than Red" mode of thinking. Here again we find an underlying assumption that communism is socially, economically, and ideologically inevitable—that once the military struggle is lost, all is lost. Not only are these assumptions patently ridiculous; they also reveal a profound misconception about the nature of social systems. The structure of a society is not determined merely by a belief. It cannot be maintained if it does not work—that is, if no one, not even those in power, is benefiting from it. How many times in history have less civilized nations conquered more civilized ones only to be entirely transformed by the cultural influence of their victims? Do we then feel less civilized than the Soviet Union? Is our system so brittle and theirs so enduring?

Actually, quite the contrary seems to be the case. For while democracy seems to be on a fairly sturdy basis in the United States (despite the efforts of self-appointed vigilantes to subvert it), there is considerable evidence that autocracy is beginning to decay in the Soviet Union.

Soviet Drift

Most Americans have great difficulty in evaluating the facts when they are confronted with evidence of decentralization in the Soviet Union, of relaxation of repressive controls, or of greater tolerance for criticism. We do not seem to sense the contradiction when we say that these changes were made in response to public discontent. For have we not also believed that an authoritarian regime, if efficiently run, can get away with ignoring the public's clamor?

There is a secret belief among us that either Khrushchev must have been mad to relax his grip or that it is all part of a secret plot to throw the West off guard: a plot too clever for naive Americans to fathom. It is seldom suggested that "de-Stalinization" took place because the rigid, repressive authoritarianism of the Stalin era was inefficient and that many additional relaxations will be forced upon the Soviet Union by the necessity of remaining amenable to technological innovation.

But the inevitable Soviet drift toward a more democratic structure is not dependent on the realism of leaders. Leaders come from communities and families, and their patterns of thought are shaped by their experiences with authority in early life, as well as by their sense of what the traffic will bear. We saw that the roots of democracy in the United States were to be found in the nature of the American family. What does the Soviet family tell us in this respect?

Pessimism regarding the ultimate destiny of Soviet political life has always been based on the seemingly fathomless capacity of the Soviet people for authoritarian submission. Their tolerance for autocratic rulers was only matched by their autocratic family system, which, in its demand for filial obedience, was equal to those of Germany, China, and many Latin countries. Acceptance of authoritarian rule was based on this early experience in the family.

But modern revolutionary movements, both fascist and communist, have tended to regard the family with some suspicion, as the preserver of old ways and as a possible refuge from the State. Fascist dictators have extolled the conservatism of the family but tended at times to set up competitive loyalties for the young. Communist revolutionaries, on the other hand, have more unambivalently attacked family loyalty as reactionary and have deliberately undermined familial allegiances, partly to increase loyalty to the state, partly to facilitate industrialization and modernization by discrediting traditional mores.

Such destruction of authoritarian family patterns is a two-edged sword that eventually cuts away political as well as familial autocracy. The state may attempt to train submission in its own youth organizations, but so long as the family remains an institution, this earlier and more enduring experience will outweigh all others. And if the family has been forced by the state to be less authoritarian, the result is obvious.

In creating a youth that has a knowledge, a familiarity, and a set of attitudes more appropriate for successful living in the changing culture than those of its parents, the autocratic state has created a Frankensteinian monster that will eventually sweep away the authoritarianism in which it is founded. The Soviet Union's attempts during the late 1930s to reverse its stand on the family perhaps reflect some realization of this fact. Khrushchev's denunciations of certain Soviet artists and intellectuals also reflect fear of a process going further than what was originally intended.

A similar ambivalence has appeared in China, where the unforeseen consequences of the slogan "all for the children" recently produced a rash of articles stressing filial obligations. As W. J. Goode points out, "The propaganda campaign against the power of the elders may lead to misunderstanding on the part of the young, who may at times abandon their filial responsibilities to the State."[7]

Further, what the derogation of parental wisdom and authority has begun, the fierce drive for technological modernization will finish. Each generation of youth will be better adapted to the changing society than its parents were. And each generation of parents will feel increasingly modest and doubtful about overvaluing its wisdom and superiority as it recognizes the brevity of its usefulness.

We cannot, of course, predict what forms democratization might take in any nation of the world, nor should we become unduly optimistic about its impact on international relations. Although our thesis predicts the democratization of the entire globe, this is a view so long range as to be academic. There are infinite opportunities for global extermination before we reach any such stage of development.

We should expect that in the earlier stages of industrialization, dictatorial regimes will prevail in all of the less developed nations. And as we well know, autocracy is still highly compatible with a lethal if short-run military efficiency. We may expect many political grotesques, some of them dangerous in the extreme, to emerge during this long period of transition, as one society after another attempts to crowd the most momentous social changes into a generation or two, working from the most varied structural baselines.

But barring some sudden decline in the rate of technological change and on the (outrageous) assumption that war will somehow be eliminated during the next half-century, it is possible to predict that after this time, democracy will be universal. Each revolutionary autocracy, as it reshuffles the family structure and pushes toward industrialization, will sow the seeds of its own destruction, and democratization will gradually engulf it.

We might, of course, rue the day. A world of mass democracies may well prove homogenized and ugly. It is perhaps beyond human social capacity to maximize both equality and understanding on the one hand, diversity on the other. Faced with this dilemma, however, many people are willing to sacrifice quaintness to social justice, and we might conclude by remarking that just as Marx, in proclaiming the inevitability of communism, did not hesitate to give some assistance to the wheels of fate, so our thesis that democracy represents the social system of the electronic era should not bar these persons from giving a little push here and there to the inevitable.

Notes

1. W. G. Bennis, "Towards a 'Truly' Scientific Management: The Concept of Organizational Health," *General Systems Yearbook,* December 1962, p. 273.

2. N. Sanford, "Social Science and Social Reform," Presidential Address for the Society for the Psychological Study of Social Issues at Annual Meeting of the American Psychological Association, Washington, D.C., August 28, 1958.

3. *Relativity for the Millions* (New York: The Macmillan Company, 1962), p. 11.

4. For a fuller discussion of this trend, see Theodore Levitt, "Marketing Myopia," HBR July–August 1960, p. 45.

5. *The Uprooted* (Boston: Little, Brown and Company, 1951).

6. W. J. Goode, *World Revolution and Family Patterns* (New York: Free Press, 1963), p. 355.

7. Ibid., pp. 313–15.

3

The Wallenda Factor

"The Wallenda Factor" is as much about life as about management. In this short essay, based on a study of 90 leaders, I tried to home in on the characteristic that allows a few people to do extraordinary things. I found that those who successfully walked the high wire in whatever field they chose had much in common. They took risks without dwelling too much on the downside. They saw error, not in terms of failure, but in terms of growth. I'm convinced that the late Karl Wallenda found in the tight-rope the answer to a question all of us should ask ourselves: "When do I feel most alive?"

One of the most impressive and memorable qualities of the leaders I studied was the way they responded to failure. Like Karl Wallenda, the great tight-rope aerialist, who once said, "The only time I feel truly alive is when I walk the tight-rope," these leaders put all their energies into their task. They simply don't think about failure, don't even use the word, relying on such synonyms as *mistake* or *glitch* or *bungle* or countless others, such as *false start, mess, hash, bollix,* or *error.* Never *failure.* One of them said during the course of an interview that "a mistake is just another way of doing things." Another said, "I try to make as many mistakes as quickly as I can in order to learn."

Shortly after Wallenda fell to his death in 1978 (traversing a 75-foot-high tight-rope in downtown San Juan, Puerto Rico), his wife, also an aerialist, discussed that fateful San Juan walk, "perhaps his most dangerous." She recalled, "All Karl thought about for three straight months prior to it was *falling.* It was the first time he'd *ever* thought about that, and it seemed to me that he put all his energy into *not falling,* not into walking the tight-rope."

Mrs. Wallenda went on to say that her husband even went so far as to personally supervise the installation of the tight-rope, making certain that the guy-wires were secure, "something he had never even thought of before."

From what I learned from my interviews with successful leaders, I can say that when Karl Wallenda poured his energies into *not falling* rather than walking the tight-rope, he was virtually destined to fail. Indeed, I call that peculiar combination of vision, persistence, consistency, and self-confidence necessary for successful tight-rope walking—the combination I found in so many leaders—the Wallenda Factor.

An example of the Wallenda Factor comes in an interview with Fletcher Byrom, who retired recently from the presidency of the Koppers Company, a diversified engineering, construction, and chemicals company. When asked about the "hardest decision he ever had to make," he said, "I don't know what a hard decision is. I may be a strange animal, but I don't worry. Whenever I make a decision, I start out recognizing there's a strong likelihood I'm going to be wrong. All I can do is the best I can. To worry puts obstacles in the way of clear thinking."

Or consider Ray Meyer—perhaps the winningest coach in college basketball—who led DePaul University to 40 straight years of winning seasons. When his team dropped its first home game after 29 straight home-court victories, I asked him how he felt about it. His response was vintage Wallenda: "Great! Now we can start concentrating on winning, *not* on not losing." And then there is Harold Prince, the Broadway producer. He regularly calls a press conference the morning after one of his Broadway plays opens—before reading any reviews—in order to announce plans for his *next* play.

Effective leaders overlook error and constantly embrace positive goals. They pour their energies into the task, not into looking behind and dredging up excuses for past events. For a lot of people, the word "failure" carries with it a finality, the absence of movement characteristic of a dead thing, to which the automatic human reaction is helpless discouragement. But, for the successful leader, failure is a beginning, the springboard of hope.

One CEO I interviewed, James Rouse, the famed city planner and developer, said that when he was dissatisfied with the looks of some housing in his Columbia, Maryland, project, he tried to influence the next design by nagging and correcting his team of architects. He got nowhere. Then he decided to stop correcting them and tried to influence them by demonstrating what he wanted, *what he was for.* Inspired by Rouse's vision, the architects went on to create some of the most eye-catching and functional housing in the country. What this illustrates is that the self-confidence of leaders is *contagious.* Two more examples: In the early days of Polaroid, Edwin Land continually inspired his team to "achieve the impossible." Land's compelling self-confidence convinced his managers and researchers that they couldn't fail. When William Hewitt took over Deere and Company in the mid-1950's, he turned a sleepy, old-line farm implements firm into a leader among modern multinational corporations. His secret? Commitment. Confidence. Vision. And always asking, "Can't we do this a little better?" And the employees rose to the occasion. As one long-time Deere employee put it: "Hewitt made us learn how good we were." Because they know where they are going, great leaders inspire the people who work for them so that they, too, can walk the tight-rope. That is one of the reasons why organizations run by great leaders often appear so productive.

Although leading is a "job" for which leaders are handsomely paid, where their rewards come from—and what they truly value—is a sense of adventure and play. In my interviews, they describe work in ways that scientists and other creative types use: "exploring a new space," "solving a problem," "designing or discovering something new." Like explorers, scientists, and artists, they seem to focus their attention on a limited field—their task—forget personal problems, lose their sense of time, feel competent and in control. When these elements are present, leaders truly enjoy what they're doing and stop worrying about whether the activity will be productive or not, whether their activities will be rewarded or not, whether what they are doing will work or not. They are walking the tight-rope.

I've wondered, from time to time, if this fusion of work and play, where, quoting from a Robert Frost poem, "love and need are one," is a positive addiction. My guess is that it is a healthy addiction, not only for individual leaders but for society. Great leaders are like the Zen archer who develops his skills to the point where the desire to hit the target becomes extinguished, and man, arrow, and target become indivisible components of the same process. That's good for the leaders. And when this style of influence works to attract and empower people to join them on the tight-rope, that's good for organizations and for society. Hail the Wallenda Factor!

4

The Coming Death of Bureaucracy

Originally published in 1966, this piece argues that bureaucracy is an institution whose time has past. The essay falters in some of its particulars—certainly General Motors is not the formidable empire it was twenty-five years ago. But the basic premise is truer today than it ever was. The wisdom of "destroying the pyramids," as SAS's Jan Carlzon calls it, has become a First Principle of successful organizations as disparate as Jack Welch's General Electric and Percy Barnevik's Asea Brown Boveri.

Not far from the new Government Center in downtown Boston, a foreign visitor walked up to a sailor and asked why American ships were built to last only a short time. According to the tourist, "The sailor answered without hesitation that the art of navigation is making such rapid progress that the finest ship would become obsolete if it lasted beyond a few years. In these words which fell accidentally from an uneducated man, I began to recognize the general and systematic idea upon which your great people direct all their concerns."

The foreign visitor was that shrewd observer of American morals and manners, Alexis de Tocqueville, and the year was 1835. He would not recognize Scollay Square today. But he had caught the central theme of our country: its preoccupation, its *obsession* with change. One thing is, however, new since de Tocqueville's time: the *acceleration* of newness, the changing scale and scope of change itself. As Dr. Robert Oppenheimer said, ". . . the world alters as we walk in it, so that the years of man's life measure not some small growth or rearrangement or moderation of what was learned in childhood, but a great upheaval."

How will these accelerating changes in our society influence human organizations?

A short while ago, I predicted that we would, in the next 25 to 50 years, participate in the end of bureaucracy as we know it and in the rise of new social systems better suited to the twentieth-century demands of industrialization. This forecast was based on the evolutionary principle that every age develops an organizational form appropriate to its genius, and that the prevailing form, known by sociologists as bureaucracy and by most businessmen as "damn bureaucracy," was out of joint with contemporary realities. I realize now that my distant prophecy is already a distinct reality so that prediction is already foreshadowed by practice.

I should like to make clear that by bureaucracy I mean a chain of command structured on the lines of a pyramid—the typical structure which coordinates the business of almost every human organization we know of: industrial, governmental, of universities and research and development laboratories, military, religious, voluntary. I do not have in mind those fantasies so often dreamed up to describe complex organizations. These fantasies can be summarized in two grotesque stereotypes. The first I call "Organization as Inkblot"—an actor steals around an uncharted wasteland, growing more restive and paranoid by the hour, while he awaits orders that never come. The other specter is "Organization as Big Daddy"—the actors are square people plugged into square holes by some omniscient and omnipotent genius who can cradle in his arms the entire destiny of man by way of computer and TV. Whatever the first image owes to Kafka, the second owes to George Orwell's *1984*.

Bureaucracy, as I refer to it here, is a useful social invention that was perfected during the industrial revolution to organize and direct the activities of a business firm. Most students of organizations would say that its anatomy consists of the following components:

1. A well-defined chain of command.

2. A system of procedures and rules for dealing with all contingencies relating to work activities.

3. A division of labor based on specialization.

4. Promotion and selection based on technical competence.

5. Impersonality in human relations.

It is the pyramid arrangement we see on most organizational charts.

The bureaucratic "machine model" was developed as a reaction against the personal subjugation, nepotism and cruelty, and the capricious and subjective judgments which passed for managerial practices during the early days of the industrial revolution. Bureaucracy emerged out of the organizations' need for order and precision and the workers' demands for impartial treatment. It was an organization ideally suited to the values and demands of the Victorian era. And just as bureaucracy emerged as a creative response to a radically new age, so today new organizational shapes are surfacing before our eyes.

First I shall try to show why the conditions of our modern industrial world will bring about the death of bureaucracy. In the second part of this article I will suggest a rough model of the organization of the future.

Four Threats

There are at least four relevant threats to bureaucracy:

1. Rapid and unexpected change.

2. Growth in size where the volume of an organization's traditional activities is not enough to sustain growth. (A number of factors are included here, among them: bureaucratic overhead; tighter controls and impersonality due to bureaucratic sprawls; outmoded rules and organizational structures.)

3. Complexity of modern technology where integration between activities and persons of very diverse, highly specialized competence is required.

4. A basically psychological threat springing from a change in managerial behavior.

It might be useful to examine the extent to which these conditions exist *right now:*

Rapid and Unexpected Change

Bureaucracy's strength is its capacity to efficiently manage the routine and predictable in human affairs. It is almost enough to cite the knowledge and population explosion to raise doubts about its contemporary viability. More revealing, however, are the statistics which demonstrate these overworked phrases:

a. Our productivity output per man hour may now be doubling almost every 20 years rather than every 40 years, as it did before World War II.

b. The Federal Government alone spent $16 billion in research and development activities in 1965; it will spend $35 billion by 1980.

c. The time lag between a technical discovery and recognition of its commercial uses was: 30 years before World War I, 16 years between the Wars, and only 9 years since World War II.

d. In 1946, only 42 cities in the world had populations of more than one million. Today there are 90. In 1930, there were 40 people for each square mile of the earth's land surface. Today there are 63. By 2000, it is expected, the figure will have soared to 142.

Bureaucracy, with its nicely defined chain of command, its rules and its rigidities, is ill-adapted to the rapid change the environment now demands.

Growth in Size

While, in theory, there may be no natural limit to the height of a bureaucratic pyramid, in practice the element of complexity is almost invariably introduced with great size. International operation, to cite one significant new element, is the rule rather than exception for most of our biggest corporations. Firms like Standard Oil Company (New Jersey) with over 100 foreign affiliates, Mobil Oil Corporation, The National Cash Register Company, Singer Company, Burroughs Corporation and Colgate-Palmolive

Company derive more than half their income or earnings from foreign sales. Many others—such as Eastman Kodak Company, Chas. Pfizer & Company, Inc., Caterpillar Tractor Company, International Harvester Company, Corn Products Company and Minnesota Mining & Manufacturing Company—make from 30 to 50 percent of their sales abroad. General Motors Corporation sales are not only nine times those of Volkswagen, they are also bigger than the Gross National Product of the Netherlands and well over the GNP of a hundred other countries. If we have seen the sun set on the British Empire, we may never see it set on the empires of General Motors, ITT, Shell and Unilever.

Labor Boom

Increasing Diversity

Today's activities require persons of very diverse, highly specialized competence.

Numerous dramatic examples can be drawn from studies of labor markets and job mobility. At some point during the past decade, the U.S. became the first nation in the world ever to employ more people in service occupations than in the production of tangible goods. Examples of this trend:

a. In the field of education, the *increase* in employment between 1950 and 1960 was greater than the total number employed in the steel, copper and aluminum industries.

b. In the field of health, the *increase* in employment between 1950 and 1960 was greater than the total number employed in automobile manufacturing in either year.

c. In financial firms, the *increase* in employment between 1950 and 1960 was greater than total employment in mining in 1960.

These changes, plus many more that are harder to demonstrate statistically, break down the old, industrial trend toward more and more people doing either simple or undifferentiated chores.

Hurried growth, rapid change and increase in specialization—pit these three factors against the five components of the pyramid structure described on pages 62–63, and we should expect the pyramid of bureaucracy to begin crumbling.

Change in Managerial Behavior

There is, I believe, a subtle but perceptible change in the philosophy underlying management behavior. Its magnitude, nature and antecedents, however, are shadowy because of the difficulty of assigning numbers. (Whatever else statistics do for us, they most certainly provide a welcome illusion of certainty.) Nevertheless, real change seems underway because of:

a. A new concept of *man,* based on increased knowledge of his complex and shifting needs, which replaces an oversimplified, innocent, pushbutton idea of man.

b. A new concept of *power,* based on collaboration and reason, which replaces a model of power based on coercion and threat.

c. A new concept of *organizational values,* based on humanistic-democratic ideals, which replaces the depersonalized mechanistic value system of bureaucracy.

The primary cause of this shift in management philosophy stems not from the bookshelf but from the manager himself. Many of the behavioral scientists, like Douglas McGregor or Rensis Likert, have clarified and articulated—even legitimized—what managers have only half registered to themselves. I am convinced, for example, that the popularity of McGregor's book, *The Human Side of Enterprise,* was based on his rare empathy for a vast audience of managers who are wistful for an alternative to the mechanistic concept of authority, i.e., that he outlined a vivid utopia of more authentic human relationships than most organizational practices today allow. Furthermore, I suspect that the desire for relationships in business has little to do with a profit motive per se, though it is often rationalized as doing so. The real push for these changes stems from the need, not only to humanize the organization, but to use it as a crucible of personal growth and the development of self-realization.[1]

The core problems confronting any organization fall, I believe, into five major categories. First, let us consider the problems, then let us see how our twentieth-century conditions of constant change have made the bureaucratic approach to these problems obsolete.

Integration

The problem is how to integrate individual needs and management goals. In other words, it is the inescapable conflict between individual needs (like "spending time with the family") and organizational demands (like meeting deadlines).

Under twentieth-century conditions of constant change there has been an emergence of human sciences and a deeper understanding of man's complexity. Today, integration encompasses the entire range of issues concerned with incentives, rewards and motivations of the individual, and how the organization succeeds or fails in adjusting to these issues. In our society, where personal attachments play an important role, the individual is appreciated, and there is genuine concern for his well-being, not just in a veterinary-hygiene sense, but as a moral, integrated personality.

Paradoxical Twins

The problem of integration, like most human problems, has a venerable past. The modern version goes back at least 160 years and was precipitated by an historical paradox: the twin births of modern individualism and modern industrialism. The former brought about a deep concern for and a passionate interest in the individual and his personal rights. The latter brought about increased mechanization of organized activity. Competition between the two has intensified as each decade promises more freedom and hope for man and more stunning achievements for technology. I believe that our society *has* opted for more humanistic and democratic values, however unfulfilled they may be in practice. It will "buy" these values even at loss in efficiency because it feels it can now afford the loss.

Social Influence

This problem is essentially one of power and how power is distributed. It is a complex issue and alive with controversy, partly because studies of leadership and power distribution can be interpreted in many ways, and almost always in ways which coincide with one's biases (including a cultural leaning toward democracy).

The problem of power has to be seriously reconsidered because of dramatic situational changes which make the possibility of one-man rule not necessarily "bad" but impractical. I refer to changes in top management's role.

Peter Drucker, over 12 years ago, listed 41 major responsibilities of the chief executive and declared that "90 percent of the trouble we are having with the chief executive's job is rooted in our superstition of the one-man chief." Many factors make one-man control obsolete, among them: the broadening product base of industry; impact of new technology; the scope of international operation; the separation of management from ownership; the rise of trade unions and general education. The real power of the "chief" has been eroding in most organizations even though both he and the organization cling to the older concept.

Collaboration

This is the problem of managing and resolving conflicts. Bureaucratically, it grows out of the very same process of conflict and stereotyping that has divided nations and communities. As organizations become more complex, they fragment and divide, building tribal patterns and symbolic codes which often work to exclude others (secrets and jargon, for example) and on occasion to exploit differences for inward (and always fragile) harmony.

Recent research is shedding new light on the problem of conflict. Psychologist Robert R. Blake in his stunning experiments has shown how simple it is to induce conflict, how difficult to arrest it. Take two groups of people who have never before been together, and give them a task which will be judged by an impartial jury. In less than an hour, each group devolves into a tightly-knit band with all the symptoms of an "in group." They regard their product as a "master-work" and the other group's as "commonplace" at best. "Other" becomes "enemy." "We are good, they are bad; we are right, they are wrong."

Rabbie's Reds and Greens

Jaap Rabbie, conducting experiments on intergroup conflict at the University of Utrecht, has been amazed by the ease with which conflict and stereotype develop. He brings into an experimental room two groups and distributes green name tags and pens to one group, red pens and tags to the other. The two groups do not compete; they do not even interact. They are only in sight of each other while they silently complete a questionnaire. Only ten minutes are needed to activate defensiveness and fear, reflected in the hostile and irrational perceptions of both "reds" and "greens."

Adaptation

This problem is caused by our turbulent environment. The pyramid structure of bureaucracy, where power is concentrated at the top, seems the perfect way to "run a railroad." And for the routine tasks of the nineteenth and early twentieth centuries, bureaucracy was (in some respects it still is) a suitable social arrangement. However, rather than a placid and predictable environment, what predominates today is a dynamic and uncertain one where there is deepening interdependence among economic, scientific, educational, social and political factors in the society.

Revitalization

This is the problem of growth and decay. As Alfred North Whitehead has said: "The art of free society consists first in the maintenance of the symbolic code, and secondly, in the fearlessness of revision. . . . Those societies which cannot combine reverence to their symbols with freedom of revision must ultimately decay. . . ."

Growth and decay emerge as the penultimate conditions of contemporary society. Organizations, as well as societies, must be concerned with those social structures that engender buoyancy, resilience and a "fearlessness of revision."

I introduce the term "revitalization" to embrace all the social mechanisms that stagnate and regenerate, as well as the process of this cycle. The elements of revitalization are:

1. An ability to learn from experience and to codify, store and retrieve the relevant knowledge.

2. An ability to "learn how to learn," that is, to develop methods for improving the learning process.

3. An ability to acquire and use feed-back mechanisms on performance, in short, to be self-analytical.

4. An ability to direct one's own destiny.

These qualities have a good deal in common with what John Gardner calls "self-renewal." For the organization, it means conscious attention to its own evolution. Without a planned methodology and explicit direction, the enterprise will not realize its potential.

Integration, distribution of power, collaboration, adaptation and *revitalization*—these are the major human problems of the next 25 years. How organizations cope with and manage these tasks will undoubtedly determine the viability of the enterprise.

Against this background I should like to set forth some of the conditions that will dictate organizational life in the next two or three decades.

The Environment

Rapid technological change and diversification will lead to more and more partnerships between government and business. It will be a truly mixed economy. Because of the immensity and expense of the projects, there will be fewer identical units competing in the same markets and organizations will become more interdependent.

The four main features of this environment are:

a. Interdependence rather than competition.

b. Turbulence and uncertainty rather than readiness and certainty.

c. Large-scale rather than small-scale enterprises.

d. Complex and multinational rather than simple national enterprises.

"Nice"—and Necessary

Population Characteristics

The most distinctive characteristic of our society is education. It will become even more so. Within 15 years, two-thirds of our population living in metropolitan areas will have attended college. Adult education is growing even faster, probably because of the rate of professional obsolescence. The Killian report showed that the average engineer required further education only ten years after getting his degree. It will be almost routine for the experienced physician, engineer and executive to go back to school for advanced training every two or three years. All of this education is not just "nice." It is necessary.

One other characteristic of the population which will aid our understanding of organizations of the future is increasing job mobility. The ease of transportation, coupled with the needs of a dynamic environment, change drastically the idea of "owning" a job—or "having roots." Already 20 percent of our population change their mailing address at least once a year.

Work Values

The increased level of education and mobility will change the values we place on work. People will be more intellectually committed to their jobs and will probably require more involvement, participation and autonomy.

Also, people will be more "other-oriented," taking cues for their norms and values from their immediate environment rather than tradition.

Tasks and Goals

The tasks of the organization will be more technical, complicated and unprogrammed. They will rely on intellect instead of muscle. And they will be too complicated for one person to comprehend, to say nothing of control. Essentially, they will call for the collaboration of specialists in a project or a team-form of organization.

There will be a complication of goals. Business will increasingly concern itself with its adaptive or innovative-creative capacity. In addition, supragoals will have to be articulated, goals which shape and provide the foundation for the goal structure. For example, one might be a system for detecting new and changing goals; another could be a system for deciding priorities among goals.

Finally, there will be more conflict and contradiction among diverse standards for organizational effectiveness. This is because professionals tend to identify more with the goals of their profession than with those of their immediate employer. University professors can be used as a case in point. Their inside work may be a conflict between teaching and research, while more of their income is derived from outside sources, such as foundations and consultant work. They tend not to be good "company men" because they divide their loyalty between their professional values and organizational goals.

Key Word: "Temporary"

Organization

The social structure of organizations of the future will have some unique characteristics. The key word will be "temporary." There will be adaptive, rapidly changing *temporary* systems. These will be task forces organized around problems to be solved by groups of relative strangers with diverse professional skills. The groups will be arranged on an organic rather than mechanical model; they will evolve in response to a problem rather than to programmed role expectations. The executive thus becomes a coordinator or "linking pin" between various task forces. He must be a man who can speak the polyglot jargon of research, with skills to relay information and to mediate between groups. People will be evaluated not vertically according to rank and status, but flexibly and functionally according to skill and professional training. Organizational charts will consist of project groups rather than stratified functional groups. (This trend is already visible in the aerospace and construction industries, as well as many professional and consulting firms.)

Adaptive, problem-solving, temporary systems of diverse specialists, linked together by coordinating and task-evaluating executive specialists in an organic flux—this is the organization form that will gradually replace bureaucracy as we know it. As no catchy phrase comes to mind, I call this an organic-adaptive structure. Organizational arrangements of this sort may not only reduce the intergroup conflicts mentioned earlier; they may also induce honest-to-goodness creative collaboration.

Motivation

The organic-adaptive structure should increase motivation and thereby effectiveness, because it enhances satisfactions intrinsic to the task. There is a harmony between the educated individual's need for tasks that are meaningful, satisfactory and creative and a flexible organizational structure.

I think that the future I describe is not necessarily a "happy" one. Coping with rapid change, living in temporary work systems, developing meaningful relations and then breaking them—all augur social strains and psychological tensions. Teaching how to live with ambiguity, to identify with the adaptive process, to make a virtue out of contingency, and to be self-directing—these will be the tasks of education, the goals of maturity, and the achievement of the successful individual.

No Delightful Marriages

In these new organizations of the future, participants will be called upon to use their minds more than at any other time in history. Fantasy, imagination and creativity will be legitimate in ways that today seem strange. Social structures will no longer be instruments of psychic repression but will increasingly promote play and freedom on behalf of curiosity and thought.

One final word: While I forecast the structure and value coordinates for organizations of the future and contend that they are inevitable, this should not bar any of us from giving the inevitable a little push. The French moralist may be right in saying that there are no delightful marriages, just good ones. It is possible that if managers and scientists continue to get their heads together in organizational revitalization, they *might* develop delightful organizations—just possibly.

I started with a quote from de Tocqueville and I think it would be fitting to end with one: "I am tempted to believe that what we call necessary institutions are often no more than institutions to which we have grown accustomed. In matters of social constitution, the field of possibilities is much more extensive than men living in their various societies are ready to imagine."

Note

1. Let me propose an hypothesis to explain this tendency. It rests on the assumption that man has a basic need for transcendental experiences, somewhat like the psychological rewards which William James claimed religion provided—"an assurance of safety and a temper of peace, and in relation to others, a preponderance of living affections." Can it be that as religion has become secularized, less transcendental, men search for substitutes such as close interpersonal relationships, psychoanalysis—even the release provided by drugs such as LSD?

5

The Four Competencies of Leadership

Much of what I've learned in decades of studying leadership was first distilled in this essay from 1984. The standard criteria for choosing top-level managers are technical competence, people skills, conceptual skills, judgment, and character. And yet effective leadership is overwhelmingly the function of only one of these—character. (Judgment is an important secondary criterion.) If you ask subordinates what they want in a leader, they usually list three things: direction or vision, trustworthiness, and optimism. Like effective parents, lovers, teachers, and therapists, good leaders make people hopeful.

For nearly five years I have been researching a book on leadership. During this period, I have traveled around the country spending time with 90 of the most effective, successful leaders in the nation; 60 from corporations and 30 from the public sector.

My goal was to find these leaders' common traits, a task that has required much more probing than I expected. For a while, I sensed much more diversity than commonality among them. The group comprises both left-brain and right-brain thinkers; some who dress for success and some who don't; well-spoken, articulate leaders and laconic, inarticulate ones; some John Wayne types and some who are definitely the opposite. Interestingly, the group includes only a few stereotypically charismatic leaders.

Despite the diversity, which is profound and must not be underestimated, I identified certain areas of competence shared by all 90. Before presenting those findings, though, it is important to place this study in context, to review the mood and events in the United States just before and during the research.

Decline and Malaise

When I left the University of Cincinnati late in 1977, our country was experiencing what President Carter called "despair" or "malaise." From 1960 to 1980, our institutions' credibility had eroded steadily. In an article about that period entitled "Where Have All the Leaders Gone?," I described how difficult the times were for leaders, including university presidents like myself.

I argued that, because of the complexity of the times, leaders felt impotent. The assassinations of several national leaders, the Vietnam war, the Watergate scandal, the Iranian hostage crisis and other events led to a loss of trust in our institutions and leadership.

I came across a quotation in a letter Abigail Adams wrote to Thomas Jefferson in 1790: "These are the hard times in which a genius would wish to live." If, as she believed, great necessities summon great leaders, I wanted to get to know the leaders brought forth by the current malaise. In a time when bumper stickers appeared reading "Impeach Someone," I resolved to seek out leaders who were effective under these adverse conditions.

At the same time that America suffered from this leadership gap, it was suffering from a productivity gap. Consider these trends:

- In the 1960s, the average GNP growth was 4.1 percent; in the 1970s, it was 2.9 percent; in 1982, it was negative.
- The U.S. standard of living, the world's highest in 1972, now ranks fifth.
- In 1960, when the economies of Europe and Japan had been rebuilt, the U.S. accounted for 25 percent of the industrial nations' manufacturing exports and supplied 98 percent of its domestic markets. Now, the U.S. has less than a 20 percent share of the world market, and that share is declining.
- In 1960, U.S. automobiles had a 96 percent market share;

today we have about 71 percent. The same holds true for consumer electronics; in 1960 it was 94.4 percent, in 1980 only 49 percent. And that was before Sony introduced the Walkman!

In addition to leadership and productivity gaps, a subtler "commitment gap" existed, that is, a reluctance to commit to one's work or employer.

The Public Agenda's recent survey of working Americans shows the following statistics. Less than one out of four jobholders (23 percent) says he or she currently works at full potential. Nearly half say they do not put much effort into their jobs above what is required. The overwhelming majority, 75 percent, say they could be significantly more effective on their job than they are now. And nearly 6 in 10 working Americans believe that "most people do not work as hard as they used to."

A number of observers have pointed out the considerable gap between the number of hours people are paid to work and the numbers of hours they spend on productive labor. Evidence developed recently by the University of Michigan indicates the gap may be widening. They found the difference between paid hours and actual working hours grew 10 percent between 1970 and 1980.

This increasing commitment gap leads to the central question: How can we empower the work force and reap the harvest of human effort?

If I have learned anything from my research, it is this: The factor that empowers the work force and ultimately determines which organizations succeed or fail is the leadership of those organizations. When strategies, processes or cultures change, the key to improvement remains leadership.

The Sample: 90 Leaders

For my study, I wanted 90 effective leaders with proven track records. The final group contains 60 corporate executives, most, but not all, from Fortune 500 companies, and 30 from the public sector. My goal was to find people with leadership ability, in contrast to just "good managers"—true leaders who affect the culture, who are the social architects of their organizations and who create and maintain values.

Leaders are people who do the right thing; managers are people who do things right. Both roles are crucial, and they differ profoundly. I often observe people in top positions doing the wrong thing well.

Given my definition, one of the key problems facing American organizations (and probably those in much of the industrialized world) is that they are underled and overmanaged. They do not pay enough attention to doing the right thing, while they pay too much attention to doing things right. Part of the fault lies with our schools of management; we teach people how to be good technicians and good staff people, but we don't train people for leadership.

The group of 60 corporate leaders was not especially different from any profile of top leadership in America. The median age was 56. Most were white males, with six black men and six women in the group. The only surprising finding was that all the CEOs not only were married to their first spouse, but also seemed enthusiastic about the institution of marriage. Examples of the CEOS are Bill Kieschnick, chairman and CEO of Arco, and the late Ray Kroc of McDonald's restaurants.

Public-sector leaders included Harold Williams, who then chaired the SEC; Neil Armstrong, a genuine all-American hero who happened to be at the University of Cincinnati; three elected officials; two orchestra conductors; and two winning athletics coaches. I wanted conductors and coaches because I mistakenly believed they were the last leaders with complete control over their constituents.

After several years of observation and conversation, I have defined four competencies evident to some extent in every member of the group. They are:

- management of attention;
- management of meaning;
- management of trust;
- management of self.

Management of Attention

One of the traits most apparent in these leaders is their ability to draw others to them, because they have a vision, a dream, a set of intentions, an agenda, a frame of reference. They communicate an extraordinary focus of commitment, which attracts people to them. One of these leaders was described as making people want to join in with him; he enrolls them in his vision.

Leaders, then, manage attention through a compelling vision that brings others to a place they have not been before. I came to this understanding in a roundabout way, as this anecdote illustrates.

One of the people I most wanted to interview was one of the few I couldn't seem to reach. He refused to answer my letters or phone calls. I even tried getting in touch with the members of his board. He is Leon Fleischer, a well-known child prodigy who grew up to become a prominent pianist, conductor and musicologist.

When I called him originally to recruit him for the University of Cincinnati faculty, he declined and told me he was working with orthopedic specialists to regain the use of his right hand. He did visit the campus, and I was impressed with his commitment to staying in Baltimore, near the medical institution where he received therapy.

Fleischer was the only person who kept turning me down for an interview, and finally I gave up. A couple of summers later I was in Aspen, Colorado, while Fleischer was conducting the Aspen Music Festival. I tried to reach him again, even leaving a note on his dressing room door, but I got no answer.

One day in downtown Aspen, I saw two perspiring young cellists carrying their instruments and offered them a ride to the music tent. They hopped in the back of my jeep, and, as we rode, I questioned them about Fleischer.

"I'll tell you why he is so great," said one. "He doesn't waste our time."

Fleischer finally agreed not only to be interviewed but to let me watch him rehearse and conduct music classes. I linked the way I saw him work with that simple sentence, "He doesn't waste our time." Every moment Fleischer was before the orchestra, he knew exactly what sound he wanted. He didn't waste time because his intentions were always evident. What united him with the other musicians was their concern with intention and outcome.

When I reflected on my own experience, it struck me that when I was most effective, it was because I knew what I wanted. When I was ineffective, it was because I was unclear about it.

So, the first leadership competency is the management of attention through a set of intentions or a vision, not in a mystical or religious sense, but in the sense of outcome, goal or direction.

Management of Meaning

To make dreams apparent to others, and to align people with them, leaders must communicate their vision. Communication and alignment work together.

Consider, for example, the contrasting styles of Presidents Reagan and Carter. Ronald Reagan is called "the great communicator"; one of his speech writers said Reagan can read the phone book and make it interesting. The reason is that Reagan uses metaphors with which people can identify.

In his first budget message, for example, Reagan described a trillion dollars by comparing it to piling up dollar bills beside the Empire State Building. Reagan, to use one of Alexander Haig's coinages, "tangibilitated" the idea. Leaders make ideas tangible and real to others, so they can support them. For no matter how marvelous the vision, the effective leader must use a metaphor, a word or a model to make that vision clear to others.

In contrast, President Carter was boring. Carter was one of our best informed presidents; he had more facts at his finger tips than almost any other president. But he never made the meaning come through the facts.

I interviewed an assistant secretary of commerce appointed by Carter, who told me that after four years in his administration, she still did not know what Jimmy Carter stood for. She said that working for him was like looking through the wrong side of a tapestry; the scene was blurry and indistinct.

The leader's goal is not mere explanation or clarification but the creation of meaning. My favorite baseball joke is exemplary: In the ninth inning of a key playoff game, with a 3 and 2 count on the batter, the umpire hesitates a split second in calling the pitch. The batter whirls around angrily and says, "Well, what was it?" The umpire barks back, "It ain't *nothing* until *I* call it!"

The more far-flung and complex the organization, the more critical is this ability. Effective leaders can communicate ideas through several organizational layers, across great distances, even through the jamming signals of special interest groups and opponents.

When I was a university president, a group of administrators and I would hatch what we knew was a great idea. Then we would do the right thing: delegate, delegate, delegate. But when the product or policy finally appeared, it scarcely resembled our original idea.

This process occurred so often that I gave it a name: the Pinocchio Effect. (I am sure Geppetto had no idea how Pinocchio would look when he finished carving him.) The Pinocchio Effect leaves us surprised. Because of inadequate communication, results rarely resemble our expectations.

We read and hear so much about information that we tend to overlook the importance of meaning. Actually, the more bombarded a society or organization, the more deluged with facts and images, the greater its thirst for meaning. Leaders integrate facts, concepts and anecdotes into meaning for the public.

Not all the leaders in my group are word masters. They get people to understand and support their goals in a variety of ways.

The ability to manage attention and meaning comes from the whole person. It is not enough to use the right buzz word or a cute technique, or to hire a public relations person to write speeches.

Consider, instead, Frank Dale, publisher of the Los Angeles afternoon newspaper, *The Herald Examiner.* Dale's charge was to cut into the market share of his morning competitor, *The L.A.*

Times. When he first joined the newspaper a few years ago, he created a campaign with posters picturing the *Herald Examiner* behind and slightly above the *Times.* The whole campaign was based on this potent message of how the *Herald Examiner* would overtake the *Times.*

I interviewed Dale at his office, and when he sat down at his desk and fastened around him a safety belt like those on airplanes, I couldn't supress a smile. He did this to remind me and everybody else of the risks the newspaper entailed. His whole person contributed to the message.

No one is more cynical than a newspaper reporter. You can imagine the reactions that traveled the halls of the *Herald Examiner* building. At the same time, nobody forgot what Frank Dale was trying to communicate. And that is the management of meaning.

Management of Trust

Trust is essential to all organizations. The main determinant of trust is reliability, what I call constancy. When I talked to the board members or staffs of these leaders, I heard certain phrases again and again: "She is all of a piece." "Whether you like it or not, you always know where he is coming from, what he stands for."

When John Paul II visited this country, he gave a press conference. One reporter asked how the Pope could account for allocating funds to build a swimming pool at the papal summer palace. He responded quickly: "I like to swim. Next question." He did not rationalize about medical reasons or claim he got the money from a special source.

A recent study showed people would much rather follow individuals they can count on, even when they disagree with their viewpoint, than people they agree with but who shift positions frequently. I cannot emphasize enough the significance of constancy and focus.

Margaret Thatcher's reelection in Great Britain is another excellent example. When she won office in 1979, observers predicted she quickly would revert to defunct Labor Party policies. She did not. In fact, not long ago a *London Times* article appeared headlined (parodying Christopher Fry's play) "The Lady's Not for Returning." She has not turned; she has been constant, focused and all of a piece.

Management of Self

The fourth leadership competency is management of self, knowing one's skills and deploying them effectively. Management of self is critical; without it, leaders and managers can do more harm than good. Like incompetent doctors, incompetent managers can make life worse, make people sicker and less vital. (The term *iatrogenic,* by the way, refers to illness *caused* by doctors and hospitals.) Some managers give themselves heart attacks and nervous breakdowns; still worse, many are "carriers," causing their employees to be ill.

Leaders know themselves; they know their strengths and nurture them. They also have a faculty I think of as the Wallenda Factor, the ability to accept risk.

One CEO told me that if she had a knack for leadership, it was the capacity to make as many mistakes as she could as soon as possible, and thus get them out of the way. Another said that a mistake is simply "another way of doing things." These leaders learn from and use something that doesn't go well; it is not a failure but simply the next step.

When I asked Harold Williams, president of the Getty Foundation, to name the experience that most shaped him as a leader, he said it was being passed over for the presidency of Norton Simon. When it happened, he was furious and demanded reasons, most of which he considered idiotic. Finally, a friend told him that some of the reasons were valid and he should change. He did, and about a year and a half later became president.

Or consider coach Ray Meyer of DePaul University, whose team finally lost at home after winning 29 straight home games. I called him to ask how he felt. He said, "Great. Now we can start to concentrate on winning, not on *not* losing."

Consider Broadway producer Harold Prince, who calls a press conference the morning after his show opens, before reading the reviews, to announce his next play. Or Susan B. Anthony, who said, "Failure is impossible." Or Fletcher Byrum, who, after 22 years as president of Coopers, was asked about his hardest decision. He replied that he did not know what a hard decision was; that he never worried, that he accepted the possibility of being wrong. Byrum said that worry was an obstacle to clear thinking.

The Wallenda Factor is an approach to life; it goes beyond leadership and power in organizations. These leaders all have it.

Empowerment: The Effects of Leadership

Leadership can be felt throughout an organization. It gives pace and energy to the work and empowers the work force. Empowerment is the collective effect of leadership. In organizations with effective leaders, empowerment is most evident in four themes:

• *People feel significant.* Everyone feels that he or she makes a difference to the success of the organization. The difference may be small—prompt delivery of potato chips to a mom-and-pop grocery store or developing a tiny but essential part for an airplane. But where they are empowered, people feel that what they do has meaning and significance.

• *Learning and competence matter.* Leaders value learning and mastery, and so do people who work for leaders. Leaders make it clear that there is no failure, only mistakes that give us feedback and tell us what to do next.

• *People are part of a community.* Where there is leadership, there is a team, a family, a unity. Even people who do not especially like each other feel the sense of community. When Neil Armstrong talks about the Apollo explorations, he describes how a team carried out an almost unimaginably complex set of interdependent tasks. Until there were women astronauts, the men referred to this feeling as "brotherhood." I suggest they rename it "family."

• *Work is exciting.* Where there are leaders, work is stimulating, challenging, fascinating and fun. An essential ingredient in organizational leadership is pulling rather than pushing people toward a goal. A "pull" style of influence attracts and energizes people to enroll in an exciting vision of the future. It motivates through identification, rather than through rewards and punishments. Leaders articulate and embody the ideals toward which the organization strives.

People cannot be expected to enroll in just any exciting vision. Some visions and concepts have more staying power and are rooted more deeply in our human needs than others. I believe the lack of two such concepts in modern organizational life is largely responsible for the alienation and lack of meaning so many experience in their work.

One of these is the concept of quality. Modern industrial society has been oriented to quantity, providing more goods and services for everyone. Quantity is measured in money; we are a money-oriented society. Quality often is not measured at all, but is appreciated intuitively. Our response to quality is a feeling. Feelings of quality are connected intimately with our experience of meaning, beauty and value in our lives.

Closely linked to the concept of quality is that of dedication, even love, of our work. This dedication is evoked by quality and is the force that energizes high-performing systems. When we love our work, we need not be managed by hopes of reward or fears of punishment. We can create systems that facilitate our work, rather than being preoccupied with checks and controls of people who want to beat or exploit the system.

And that is what the human resources profession should care most about.

6
Managing the Dream

There is currently a buzz in management circles about "global leadership," as if that were something distinct from the effective management of the past. I'm increasingly convinced that good leadership is essentially the same on whatever scale it's carried out. Yes, the global manager has to find ways to understand and empathize with other cultures. But domestic organizations typically have multiple subcultures that can be just as mysterious as any lost tribe and that must also be mastered before the organization can work. However many area codes are involved, the effective leader understands this diversity and embraces it.

Pick up any business magazine or newspaper and you'll find the same story: pessimism about America's capacity to compete successfully in the new, spirited global economy. *The Wall Street Journal* laments, "The sudden emergence of America as the world's largest debtor, Japan as the globe's richest creditor, and the Soviet Union as its most ardent preacher of pacifism seems, to many Americans, to have turned the world upside down, raising doubts about whether America can or should lead." The *Washington Post* kicks in with "Kiss Number One Goodbye, Folks." A headline in the *International Herald Tribune* warns, "America, Europe Is Coming."

If there is reason to despair and join the handwringing and headshaking of doomsayers, it's because traditional American managers were brought up in a different time, when all they had to do was build the greatest mousetraps, and the world beat a path to their doors. "Leadership in a traditional U.S. company," says R.B. Horton, CEO of BP America, "consisted of creating a management able to cope with competitors who all played with basically the same deck of economic cards." And it was an American game. The competition was fierce but knowable. If you played your cards right, you could win.

But the game has changed and strange new rules have appeared. The deck has been shuffled and jokers have been added. Never before has American business faced so many challenges, and never before have there been so many choices about how to face those challenges. Uncertainties and complexities abound. The only thing truly predictable is unpredictability. The new chic is chaos chic. As Yogi Berra put it, "The future ain't what it used to be."

Constant change disturbs some managers—it always has, and it always will. Machiavelli said, "Change has no constituency." Well, it better have one—and soon. Forget about regaining global leadership. With only a single, short decade remaining before the 21st century, we must look now at what it's going to take simply to remain a player in the game. We can do that because the 21st century is with us now. Cultures don't turn sharply with the pages of the calendar—they evolve. By paying attention to what is changing today, we know what we must do better tomorrow.

Leaders, Not Managers

Given the nature and constancy of change and the transnational challenges facing American business leadership, the key to making the right choices will come from understanding and embodying the leadership qualities necessary to succeed in a mercurial global economy. To survive in the 21st century, we're going to need a new generation of leaders—leaders, not managers.

The distinction is an important one. Leaders conquer the context—the volatile, turbulent, ambiguous surroundings that sometimes seem to conspire against us and will surely suffocate us if we let them—while managers surrender to it. There are other differences, as well, and they are crucial:

- The manager administers; the leader innovates.
- The manager is a copy; the leader is an original.

- The manager maintains; the leader develops.
- The manager focuses on systems and structure; the leader focuses on people.
- The manager relies on control; the leader inspires trust.
- The manager has a short-range view; the leader has a long-range perspective.
- The manager asks how and when; the leader asks what and why.
- The manager has his eye on the bottom line; the leader has his eye on the horizon.
- The manager accepts the status quo; the leader challanges it.
- The manager is the classic good soldier; the leader is his own person.
- The manager does things right; the leader does the right thing.

Field Marshal Sir William Slim led the 14th British Army from 1943 to 1945 in the reconquest of Burma from the Japanese—one of the epic campaigns of World War II. He recognized the distinction between leaders and managers when he said: "Managers are necessary; leaders are essential. . . . Leadership is of the spirit, compounded of personality and vision. . . . Management is of the mind, more a matter of accurate calculation, statistics, methods, timetables and routine."

I've spent the last 10 years talking with leaders, including Jim Burke at Johnson & Johnson, John Scully at Apple, television producer Norman Lear, and close to 100 other men and women, some famous and some not. In the course of my research, I've learned something about the current crop of leaders and something about the kind of leadership that will be necessary to forge the future. While leaders come in every size, shape and disposition—short, tall, neat, sloppy, young, old, male and female—every leader I talked with shared at least one characteristic: a concern with a guiding purpose, an overarching vision. They were more than goal-directed. As Karl Wallenda said, "Walking the tightwire is living; everything else is waiting."

Leaders have a clear idea of what they want to do—personally and professionally—and the strength to persist in the face of setbacks, even failures. They know where they are going and why. Senator Howard Baker said of President Reagan, whom he served as Chief of Staff, "He knew who he was, what he believed in and where he wanted to go."

Managing the Dream

Many leaders find a metaphor that embodies and implements their vision. For Charles Darwin, the fecund metaphor was a branching tree of evolution on which he could trace the rise and fate of various species. William James viewed mental processes as a stream or river. John Locke focused on the falconer, whose release of a bird symbolized his "own emerging view of the creative process"—that is, the quest for human knowledge.

I think of it this way: *Leaders manage the dream.* All leaders have the capacity to create a compelling vision, one that takes people to a new place, and the ability to translate that vision into reality. Peter Drucker said that the first task of the leader is to define the mission. Max De Pree, former CEO of Herman Miller Inc., the Zeeland, MI, office furniture maker, put it another way in *Leadership Is an Art:* "The first responsibility of a leader is to define reality. The last is to say thank you. In between, the leader is a servant."

Managing the dream can be broken down into five parts. The first part is communicating the vision. Jung said: "A dream that is not understood remains a mere occurrence. Understood, it becomes a living experience."

Jim Burke spends 40 percent of his time communicating the Johnson & Johnson credo. More than 800 managers have attended J&J challenge meetings, where they go through the credo line by line to see what changes need to be made. Over the years some of those changes have been fundamental. But like the U.S. Constitution, the credo itself endures.

The other basic parts of managing the dream are recruiting meticulously, rewarding, retraining and reorganizing. Jan Carlzon, CEO of Scandinavian Air System (SAS), is a leader who embraces all five parts.

Carlzon's vision was to make SAS one of the five or six remaining international carriers by the year 1995. (He thinks that only five or six will be left by that time, and I think he's probably right.) To accomplish this, he developed two goals. The first was to make SAS 1 percent better in 100 different ways than its competitors. The second was to create a market niche.

Carlzon chose the business traveler—rather than college students, travel agent deals or any of a host of other possibilities—because he believed that this would be the most profitable niche. In order to attract business travelers, Carlzon had to make sure that every interaction they had with every SAS employee was rewarding. He had to endow every interaction with purpose, relevance, courtesy and caring. He estimated that there were 63,000 of these interactions each day between SAS employees and current or potential customers. He called these interactions "moments of truth."

Carlzon developed a marvelous cartoon book, *The Little Red Book,* to communicate the new SAS vision to employees. And he set up a corporate college in Copenhagen to train them. Just as important, he has "debureaucratized" the whole organization. The organization chart no longer looks like a pyramid; it looks like a set of circles, a galaxy. In fact, Carlzon's book, which is called *Moments of Truth* in English, is titled *Destroying the Pyramids* in its original Swedish.

One of those circles, one organizational segment, is the Copenhagen–New York route. All the pilots, the navigators, the engineers, the flight attendants, the baggage handlers, the reservations agents—everybody who has anything to do with the Copenhagen–New York route—are involved in a self-managed, autonomous work group with a gain-sharing plan so that they all participate in whatever profits that particular route brings in. There's also a Copenhagen-Frankfurt organizational segment, and so on. The whole corporation is structured in terms of these small, egalitarian groups.

General Electric CEO Jack Welch said: "Yesterday's idea of the boss, who became the boss because he or she knew one more fact than the person working for them, is yesterday's manager.

Tomorrow's person leads through a vision, a shared set of values, a shared objective." The single defining quality of leaders is the capacity to create and realize a vision. Yeats said, "In dreams begins responsibility." Vision is a waking dream. For leaders, the responsibility is to transform the vision into reality. By doing so, they transform their dominion, whether an airline, a motion picture, the computer industry or America itself.

Thoreau put it this way: "If one advances confidently in the direction of his dreams, and endeavors to live the life he has imagined, he will meet with a success in common hours. . . . If you have built castles in the air, your work need not be lost. It is where they should be. Now put the foundation under them."

The New Global Alliances

Jan Carlzon also illustrates one element that I believe will distinguish the vision of 21st-century leaders from the current model. His is a global vision; he is fully aware of the need for transnational networking and alliances.

Carlzon is not alone. A recent United Research Co./Harris survey of 150 CEOs of *Forbes 500* companies found that they saw the greatest opportunity and challenge for the future in the global market. In the same vein, senior-level managers polled in a Carnegie-Mellon University survey of business school alumni named competing effectively on a global basis as the most difficult management issue for the next decade.

Global interdependence is one of six pivotal forces working on the world today. (The others are technology, mergers and acquisitions, deregulation and reregulation, demographics and values, and the environment. Leadership is necessary for coping with each of these forces, but those are subjects for another time.) One of the first things the astute businessperson checks daily now is the yen-dollar ratio. Fifty percent of the property in downtown Los Angeles is owned by the Japanese.

Foreign investment in America—in real estate, finance and business—continues to escalate. But the changes aren't simply on our shores. In 1992, when Europe becomes a true Common

Market, it will contain 330 million consumers, compared with 240 million in this country.

American leaders who want to be a part of that new market are planning now. Michael Eisner of Disney has sent Robert Fitzpatrick to France to head up the new EuroDisney project. CalFed, which already has a bank in England, is preparing for the future with plans for banks in Brussels, Barcelona, Paris and Vienna. In Spain, AT&T has spent $220 million for a semiconductor plant, and General Electric has budgeted $1.7 billion for a plastics facility. Ford, Nissan, Sony and Matsushita have opened factories in or near Barcelona in the last two years.

In most cases, however, buying into Europe is prohibitively expensive. The shrewd leaders of the future are recognizing the wisdom of creating alliances with other organizations whose fates are correlated with their own. The Norwegian counterpart of Federal Express—which has 3,500 employees, one of the largest companies in Norway—is setting up a partnership with Federal Express. First Boston has linked up with Credit Suisse, forming FBCS. GE has recently set up a number of joint ventures with GE of Great Britain, meshing four product divisions. Despite the names, the companies hadn't been related. GE had considered buying its British namesake, but ultimately chose alliance rather than acquisition.

Buying in is not the choice of the Europeans themselves: Glaxco, a British pharmaceutical firm, made a deal with Hoffman-LaRoche for the distribution of Zantac, a stomach tranquilizer, and knocked SmithKline Beecham's Tagamet out of the game. Kabi Virtum, a Swedish pharmaceutical company, is looking for a partner in Japan to build a joint laboratory, in exchange for which the Japanese would get help in licensing drugs in Sweden.

And as for Jan Carlzon, when he tried and failed to buy Sabena, a rival airline, he established an alliance instead. SAS also works with an Argentine airline and with Eastern Airlines, sharing gates and connecting routes.

The global strategy is firmly rooted in Carlzon's vision for SAS. All leaders' guiding visions provide clearly marked road maps for their organizations; every member can see which direction the corporation is going. The communication of the vision generates excitement about the trip. The plans for the journey create order out of chaos, instill confidence and trust, and offer criteria for success. The group knows when it has arrived.

The critical factor for success in global joint ventures is a shared vision between the two companies. If you're not sure of your company's vision, how can you tell what the advantages of an alliance would be? You must be certain you have the right map before embarking on the journey. If you think your company's vision lacks definition, here are some questions that may help give it color and dimension:

- What is unique about us?
- What values are true priorities for the next year?
- What would make me professionally commit my mind and heart to this vision over the next five to 10 years?
- What does the world really need that our company can and should provide?
- What do I want our company to accomplish so that I will be committed, aligned and proud of my association with the institution?

Ask yourself those questions today. Your answers will be the fire that heats the forge of your company's future.

7

False Grit

This piece, written a dozen years ago, was one of the first to argue that assertiveness training is not enough. If women are to succeed, they must master the social etiquette of bureaucracy. They must learn how to manipulate the organizations that have effectively excluded them for so long. When I wrote it, I was thinking mostly about the formidable Florence Nightingale, who did everything that had to be done to lessen the misery of her lads, wounded in the Crimean War. She may not have been much of a nurse, but she was one hell of a change agent. And she's still an inspiring model in 1993.

There's a mythology of competence going around that says the way for a woman to succeed is to act like a man. One proponent of this new "man-scam" is Marcille Gray Williams, author of *The New Executive Woman: A Guide to Business Success,* who advises women to "learn to control your tears. Mary Tyler Moore may be able to get away with it, but you can't. Whatever you do, don't cry." Women in increasing numbers are enrolling in a variety of training and retraining programs which tell them that if they dress properly (dark gray and dark blue) and talk tough enough (to paraphrase John Wayne, "A woman's got to do what a woman's got to do"), they'll take another step up the ladder of success. Which explains why training programs for women (and men too) have become a booming growth industry.

What we see today are all kinds of workshops and seminars where women undergo a metaphorical sex change, where they acquire a tough-talking, no-nonsense, sink-or-swim macho philosophy. They're told to take on traits just the opposite of those Harvard psychoanalyst Dr. Helen H. Tartakoff assigns to women:

"endowments which include the capacity for mutuality as well as for maternity . . . for creativity as well as receptivity. In short," she sums up, "women's feminine heritage, as caretaker and peacemaker, contains the potential for improving the human condition."

Ironically, men are simultaneously encouraged to shed the same masculine character traits that women are trying to imitate through their own form of nonassertiveness and in sensitivity training programs. So it's O.K., even better than O.K., for old Charlie to cry in his office. How marvelous. How liberating. Women impersonate the macho male stereotype and men impersonate the countermacho stereotype of the women.

It's time to move beyond "sex differences" and "sex roles," beyond the myths of female and male impersonations, to a more sophisticated understanding of women in organizations. Instead of retraining women *as individuals* to acquire appropriate dress or assertiveness, we have to face up to the organization as a culture—as a system which governs behavior. For, according to research findings, the impact of the organization on success or failure is much greater than that of personality characteristics—or, for that matter, sex differences.

This realization avoids the "blame the victim" approach which explains executive success in terms of individual dispositions (whether created by temperament or socialization). The villains of the piece turn out to be complex organizations, whose power structures and avenues for opportunity routinely disadvantage those people not particularly sophisticated about how such organizations work. More often than not, those people are women, since they tend to have had less experience in learning the ropes of organizational life. This perspective suggests a different kind of strategy for the elimination of sex discrimination than the "sex roles" school of thought. Instead of retraining women (or men, for that matter) and trapping all concerned in a false dream, it's necessary to take a look at the very nature of complex organizations. It is these systems and the roles within them that women must understand. And it is, at bottom, these complex organizations which should bear the burden of change, not the women subjected to weekend bashes where male-chauvinist Pygmalion games are played to the tune of "Why Can't a Woman Be More Like a Man?"

Alfred North Whitehead cautioned us wisely: "Seek simplicity and then distrust it." To put it kindly, the trouble with too many sex-difference, sex-role training programs is that they seek simplicity but forget to distrust it. And no wonder. Simplicity is easier. It's easier to transform individuals than to transform creaky, complex systems with their bureaucratic sludge and impenetrable webs of self-interest. It's a lot easier to change an Eliza Doolittle than Victorian England's class structure. The trouble is: When Eliza returns to her old habitat in Covent Garden, the old familiar behaviors return almost immediately, and everything she learned from Professor Higgins is extinguished in days. This "fade out" effect has occurred wherever individuals are trained or re-educated outside the organizational context. What's easier can be dangerously off target.

When I discussed this with Boris Yavitz, Dean of Columbia University's Graduate School of Business, he told me, "What I fear is that women will try to take on the attributes of men in a wrong-headed attempt to disprove the old stereotypes." The women he sees in his program have all the intellectual equipment necessary for success in business; they're motivated, directed, purposeful. "We never set out eight years ago to bring in women by making it easy," says Yavitz, "and yet in the last eight years our female enrollment has increased from 5 percent to 40 percent of the school. We are holding exactly the same standards we have always held, and the women are doing superbly." As they are, by the way, in all top graduate schools of management. M.I.T. accepts from 25 percent to 40 percent, as do Harvard, Stanford, Chicago and Wharton, the bastions of management-education excellence in this country. Women's competence is well documented. From all reports, they are capable on the job—and, Yavitz adds, "They prove their competence without the need for sporting hair on their chests."

Organizations, thankfully, are too complicated for the popular delusion of simplicity and certainty, the false-grit tunnel vision of a John Wayne. The fact is that there is no one set of rules, of pro-grammed behavior, dress or skills that can apply to women or men in their attempts to succeed. Perhaps the most convincing documentation of this point is a study of 1,800 successful man-agers recently completed by The American Management Associ-ations (A.M.A.). From this study, a profile emerges: Effective managers are social initiators; they anticipate problems and pos-sible solutions. They build alliances, bring people together, devel-op networks. Their competencies cluster in several areas; *social-emotional maturity* (composed of such traits as self-control, spontaneity, perceptual objectivity, accurate self-assessment, stamina, adaptability); *entrepreneurial abilities* (efficiency, pro-ductivity); *intellectual abilities* (logical thought, conceptual ability, the diagnostic use of ideas and memory); and *interpersonal abili-ties* (self-presentation, interest in the development of others, con-cern with impact, oral communication skills, the use of socialized power, and concern with relationships).

The A.M.A. study is, without question, some of the most com-plete, systematic research ever undertaken on the attributes of the good manager. I see nothing in its findings that would give men or women (with whatever "natural endowments" one attributes to sex roles) an edge. I would also wager that most astute observers of the managerial landscape would agree with the study. Yavitz, for one, describes the effective manager as possessing "the ability for true communication—I don't mean the glib view, that you're com-municating when you make a great pitch." He insists that two-way communication is imperative: "A manager must be perceptive, must understand what she's hearing, and then be able to convey the ideas clearly to others . . . must be flexible enough to ac-knowledge that there are competing constituencies and must be sensitive enough to listen to the emotion and spirit behind the words as well as to the content . . . and she must be able to syn-thesize what she's heard, to put together something as close to the optimal solution as possible, something that makes sense.

"The manager must be able to persuade, explain, convince oth-ers why this solution is more sensible and beneficial for the whole cluster of constituencies than another solution. I surmise," he concludes, "that a high sense of responsibility and commitment, ability to cope with ambiguity, and a continuing sense of curiosity and willingness to learn are critical attributes for the successful manager."

Does either sex have a monopoly on the constellation of traits identified by the A.M.A. research or by Dean Yavitz?

A better explanation of success, it seems to me, is that those who are favorably placed in organizational structures are more likely to be successful, independent of gender, than those less favorably placed. By "favorably placed" I mean: (1) having the support of one's subordinates, (2) having clear goals and a similarly clear path to them and (3) being empowered by the organization with appropriate means to reward and punish one's subordinates. When these conditions are present, we have what scholars refer to as "situational favorableness."

Complex organizations vary enormously. Specifically, they vary with respect to their "cultures." Some organizations are formalistic in nature, rigid, hierarchical; others are collegial, relying on agreement and consensus; while still others tend to be personalistic, concerned with the self-actualization of their employees. Within organizations, too, there can be great cultural differences. Just compare Bell Telephone's Murray Hill Labs with its international headquarters at Basking Ridge, New Jersey. It's hard to find a man without a beard or with a tie at the Bell Labs at Murray Hill, and equally hard to find a man without a tie and with a beard at Basking Ridge. The accompanying diagram contrasts the values and behavior of three types of organizational systems.

Success depends greatly on being able to diagnose the particular organizational culture within which one is embedded and to develop the flexibility to respond and initiate within that structure. There's nothing sex-related about it. All that's required are the knowledge and the personal skills that that most famous of all salesmen, Professor Harold Hill of "The Music Man," expounded: "Gotta know the territory."

From Hobbes to Freud, the special character of Western (most especially American) development has been an awareness of the heterogeneity of human experience and an accentuated con-

Three Types of Organizational Cultures

	Formalistic	*Collegial*	*Personalistic*
Basis for Decision	Direction from Authority	Discussion, agreement	Directions from within
Form of Control	Rules, laws, rewards, punishments	Interpersonal, group commitments	Actions aligned with self-concept
Source of Power	Superior	What "we" think and feel	What *I* think and feel
Desired End	Compliance	Consensus	Self-Actualization
To Be Avoided	Deviation from authoritative direction; taking risks	Failure to reach consensus	Not being "true to oneself"
Time Perspective	Future	Near Future	Now
Position Relative to Others	Hierarchical	Peer	Individual
Human Relationships	Structured	Group oriented	Individually oriented
Basis for Growth	Following the established order	Peer group membership	Acting on awareness of self

sciousness of the power of the individual to overcome or shape his circumstances. As Isaiah Berlin shows in his *Russian Thinkers,* the Russian tendency has always been for the system—always the senior partner to self-affirmation—to move toward hegemony. In the United States, that partnership has been reversed, with self-affirmation in the ascendence. There is, as

Mounier, the French political writer, has warned us, a "madness in both those who treat the world as a dream and . . . a madness in those who treat the inner life as a phantom." To apply this to systems, there are those who view organizations as mirages, with no reality except that which we give them. This is one kind of madness. The other madness is that of those, like the Russians and some *echt* Marxist thinkers, who will not deal with, or even recognize, aspects of their personality, dispositions, if you will, that stem from our inner souls, or private lives. For the sake of our collective sanity, we must recognize the validity and reality of each—by organization and personality—for without that total embrace, our perspective will be dangerously skewed.

In any case, it would be a grave error to fall into the trap of underestimating the power of organizations and conceiving of executive success as dependent on toughness or softness, assertiveness or sensitivity, masculinity or femininity. That popular delusion has already caused too much damage, both to individuals who are impersonating males and females, and to the institutions for which they work.

8

On the Leading Edge of Change

I recently met with corporate and other leaders throughout Asia. That trip convinced me that the key to competitive advantage in the 1990s will be a leader's ability to create an environment that generates intellectual capital. The day is past when an organization can thrive simply by implementing the ideas of a single leader. The effective leaders of the 1990s will be those who are best able to facilitate and orchestrate ideas, whatever their source. This is a shortened version of the original essay.

In his recent book *Adhocracy: The Power to Change,* Bob Waterman tells us that most of us are like the characters in Ibsen's play *Ghosts.* "We're controlled by ideas and norms that have outlived their usefulness, that are only ghosts but have as much influence on our behavior as they would if they were alive. The ideas of men like Henry Ford, Frederick Taylor, and Max Weber—these are the ghosts that haunt our halls of management."

Most of us grew up in organizations that were dominated by the thoughts and actions of the Fords, Taylors, and Webers, the fathers of the classic bureaucratic system. Bureaucracy was a splendid social invention in its time—the 19th century. In his deathless (and deadly) prose, the German sociologist, Max Weber, first brought to the world's attention that the bureaucratic, machine model was ideal for harnessing the manpower and resources of the Industrial Revolution. To this day, most organizations retain the macho, control-and-command mentality that is intrinsic to that increasingly threadbare mode. Indeed it is possible to capture the mindset created by that obsolete paradigm in three simple words—control, order, and predict.

Recurring Themes

Over the past dozen years, interacting with and interviewing CEOs and leaders of all kinds, I am reminded of Tolstoy's remark that all happy families are alike. Several themes appeared again and again.

• *However much the CEOs differ in experience and personal style, they constitute a prism through which the fortunes of the modern world are refracted.* These leaders are emblematic of their time, forced to deal not only with the exigencies of their own organizations but also with a new social reality. Among the broader factors that underlie all their decisions: the accelerating rate and complexity of change, the emergence of new technologies, dramatic demographic shifts, and globalization. For me, all of these are reflected in a single incident. Several years ago I invited the Dalai Lama to participate in a gathering of leaders at the University of Southern California. The living embodiment of thousands of years of Tibetan spiritual wisdom graciously declined—by fax.

• *Each of these leaders has discovered that the very culture of his organization must change* because, as constituted, that culture is more devoted to perceiving itself than to meeting new challenges.

• *Each of these individuals is a leader, not a manager.* Jack Welch, Chairman and CEO of General Electric, has predicted (correctly, I believe) that: "The world of the '90s and beyond will not belong to *managers* or those who make the numbers dance, as we used to say, or those who are conversant with all the business and jargon we use to sound smart. The world will belong to passionate, driven *leaders*—people who not only have an enormous amount of energy but who can energize those whom they lead."

• *Each of these individuals understands that management is getting people to do what needs to be done. Leadership is getting people to want to do what needs to be done.* Managers push. Leaders pull. Managers command. Leaders communicate.

- *Without exception every CEO interviewed has become the Chief Transformation Officer of his organization.* As Robert Haas, Chairman and CEO of Levi Strauss & Co., observes, change isn't easy, even for those committed to it. "It's difficult to unlearn behaviors that made us successful in the past. Speaking rather than listening. Valuing people like yourself over people of another gender or from different cultures. Doing things on your own rather than collaborating. Making the decision yourself instead of asking different people for their perspectives. There's a whole range of behaviors that were highly functional in the old hierarchical organization that are dead wrong in the flatter, more responsive organization that we're seeking to become."

John Sculley, CEO of Apple, once told me: "The old hierarchical model is no longer appropriate. The new model is global in scale, an interdependent network. So the new leader faces new tests, such as how does he lead people who don't report to him—people in other companies, in Japan or Europe, even competitors. How do you lead in this idea-intensive, interdependent-network environment? It requires a wholly different set of skills, based on ideas, people skills, and values. Traditional leaders are having a hard time explaining what's going on in the world, because they're basing their explanations on their experience with the old paradigm."

Sculley also predicted that the World War II fighter pilot (the formative experience of several corporate heads, as well as President Bush) would no longer be our principal paradigm for leaders.

The organizations of the future will be networks, clusters, cross-functional teams, temporary systems, ad hoc task forces, lattices, modules, matrices—almost anything but pyramids. We don't even know yet what to call these new configurations, but we do know that the ones that succeed will be less hierarchical and have more linkages based on common goals rather than traditional reporting relationships. It is also likely that these successful organizations will embody Rosabeth Moss Kanter's "5 F's: fast, focused, flexible, friendly, and fun."

Recently, I spoke with Alvin Toffler, the all-time change maven whose paradigm-shifting book, *Future Shock,* was published in 1970. We were trying to name an organization that exists in today's environment that was immune to change and had been stable *and* prosperous. We couldn't think of one.

The CEOs I know best understand that contemporary organizations face increasing and unfamiliar sources of competition as a result of the globalization of markets, capital, labor, and information technology. To be successful, these organizations must have flexible structures that enable them to be highly responsive to customer requirements and adaptive to changes in the competitive environment. These new organizations must be leaner, have fewer layers, and be able to engage in transnational and nontraditional alliances and mergers. And they must understand a global array of business practices, customs, and cultures.

The question all these leaders are addressing, with apparent success, is: *How do you change relatively successful organizations, which, if they continue to act today the way they acted even five years ago, will undo themselves in the future?* (Remember that 47 percent of the companies that made up the Fortune 500 in 1980 were not on the list in 1990.)

The ACE Paradigm

The CEOs are telling us that the new paradigm for success has three elements: Align, Create, and Empower, or ACE.

• *Align.* Today's leader needs to align resources, particularly human resources, creating a sense of shared objectives worthy of people's support and even dedication. Alignment has much to do with the spirit and a sense of being part of a team. Great organizations inevitably develop around a shared vision. Theodore Vail had a vision of universal telephone service that would take 50 years to bring about. Henry Ford envisioned common people, not just the wealthy, owning their own automobiles. Steven Jobs, Steven Wozniak, and their Apple cofounders saw the potential of the computer to empower people. A shared vision uplifts people's aspirations. Work becomes part of pursuing a larger purpose embodied in products and services.

- *Create.* Today's leader must create a culture where ideas come through unhampered by people who are fearful. Such leaders are committed to problem-finding, not just problem-solving. They embrace error, even failure, because they know it will teach them more than success. As Norman Lear once said to me, "Wherever I trip is where the treasure lies."

Effective leaders create adaptive, creative, learning organizations. Such organizations have the ability to identify problems, however troublesome, before they become crises. These organizations are able to rally the ideas and information necessary to solve their problems. They are not afraid to test possible solutions, perhaps by means of a pilot program. And, finally, learning organizations provide opportunities to reflect on and evaluate past actions and decisions.

- *Empower.* Empowerment involves the sense people have that they are at the center of things, rather than the periphery. In an effectively led organization, everyone feels he or she contributes to its success. Empowered individuals believe what they do has significance and meaning. Empowered people have both discretion and obligations. They live in a culture of respect where they can actually do things without getting permission first from some parent figure. Empowered organizations are characterized by trust and system-wide communication.

Whatever shape the future ultimately takes, the organizations that will succeed are those that take seriously—and sustain through action—the belief that their competitive advantage is based on the development and growth of the people in them. And the men and women who guide those organizations will be a different kind of leader than we've been used to. They will be maestros, not masters, coaches, not commanders.

Today the laurel will go to the leader who encourages healthy dissent and values those followers brave enough to say *no*. The successful leader will have not the loudest voice, but the readiest ear. His or her real genius may well lie not in personal achievements, but in unleashing other people's talent.

9

Searching for the Perfect University President

There was a time when a university presidency was a dream job, an opportunity to be a leader in a world—devoutly to be wished—where ideas were as real as the furniture. Robert Hutchins of Chicago, James Conant of Harvard, and Nicholas Murray Butler of Columbia were giants who shaped not just their own institutions but the course of intellectual life in America. They had the tenure of Supreme Court justices and almost as much influence.

Over the past twenty-five years the job has changed dramatically. Today's university president has the half-life of a spring flower. He or she is often a full-time fund-raiser and a full-time defendant in legal actions involving the campus. His or her impact on the curriculum is minimal. Instead the president must be a shrewd politician and a nimble conflict manager. The rest of the time is spent working with opinionated, often eloquent stakeholders who feel they have the right, even the responsibility, to tell you what to do. All this under the watchful eye of the press. Few executive positions involve so few degrees of freedom—which may be why so few giants seek university presidencies today and why those who accept them burn out so quickly.

Introduction

During the past twelve months more than 170 colleges and universities have chosen new presidents, including two men who failed to survive even their first year in office. As of February this year [1971], at least 112 schools were still looking for a chief executive; Harvard had finally concluded its infinitely publicized search for Nathan Pusey's replacement; and tiny Franconia College, in New Hampshire, had observed its first six months under the leadership of twenty-four-year-old Leonard Botstein, certainly the youngest, if not the best known, presidential appointee of the year.

The job pays well—from $15,000 to $50,000 in most schools—and often includes housing, a generous expense account, and a comforting, if occasionally deceiving, illusion of power. But more and more college presidents have found themselves caught squarely between the zealously defended claims of a self-protective, professionalist faculty, a restive and extravagantly idealistic student body, antiquarian state legislators, and a fusty old-guard group of trustees, alumni, and businessmen, whose generosity, nevertheless, is largely responsible for keeping open the university's doors. To deal fairly with these frequently competing interests, and see to the business of providing an education for the students, has often been more than one ordinarily civilized man could manage—though the fact has not, apparently, reduced the allure of the job for a good many ambitious educators and administrators.

What follows is the story of a college presidency dangled before Warren Bennis, a forty-five-year-old sociologist, author, and veteran administrator; of a selection process that befuddled at least one candidate and pleased practically no one; and of a decision that never came.

"Did you see the Meyerson announcement? It's awful, just awful." Seymour Knox was lamenting Martin Meyerson's decision, announced in January, 1970, to step down from the presidency of State University of New York at Buffalo in order to succeed Gaylord Harnwell as president of the University of Pennsylvania. As chairman of the Buffalo Council, Knox had been crucial in bringing Meyerson to Buffalo over the cries of a local faction determined to block the appointment of "that Jew from Berkeley." Therefore, Knox was not happy about Meyerson's decision to move on.

"I feel like a crumb-bum," he complained. "Yesterday, in Philadelphia, I ran into an old friend, Bill Day, who is now chairman of the Penn Board of Trustees. Was he gloating! God, it was awful. I felt like he'd stolen my cook."

The stolen cook was not the first analogy to come to my mind, but I knew what Knox meant. University presidents are currently somewhat harder to find and keep than competent domestic help. The numbers fluctuate, but a sizable list of colleges and universities are currently conducting presidential searches. Why are there so many openings at the top? There are many reasons, but a root cause is the altered nature of the job itself. There was a time when a university president did little more than officiate at commencements and raise funds; when his tenure was roughly equal to that of a Supreme Court Justice. Not today; in the aftermath of Berkeley, Columbia, Harvard, and Kent, a university president is a full-time crisis manager. He remains in office less than five years on the average, and is usually glad to retire. Annual turnover of university and college presidents has jumped nearly 30 percent in the last three years. In the first two months of 1970, new presidents were named at forty-two colleges, while one hundred resigned during the first six months of 1970. One analyst has compared the job unfavorably with that of a pro hockey referee. I don't think he's far off the mark. The work is rough, physically exhausting, even dangerous. University presidents may lose fewer teeth than hockey officials, but they have a startling number of heart attacks. A modern university president is expected to have practical vision, a good track record in administration, and national prominence as a scholar. He must be a good public speaker, fund-raiser, writer, analyst, friend and colleague, manipulator of power, planner, co-worker, persuader, and disciplinarian. He must have an attractive family and an indefatigable and effortlessly sociable wife. He must be a Money Man, Academic Manager, Father Figure, Public Relations Man, Political Man, and Educator. In short, as one Harvard man put it, looking toward Nathan Pusey's successor, "He must be a messiah with a good speaking voice." Or as Herman B. Wells, former president of Indiana University, said in a rather more earthy way:

"He should be born with the physical strength of a Greek athlete, the cunning of a Machiavelli, the wisdom of a Solomon, the courage of a lion, if possible. But in any case he must be born with the stomach of a goat."

During the last several years, while serving in various administrative posts at State University of New York at Buffalo, I was considered for the presidencies of at least a dozen colleges and universities. Since scrutiny works both ways, the time I spent with search committees (being examined rather like a bolt of felt, as I sometimes thought) gave me an excellent opportunity to study the process by which our universities choose presidents. Each of the twelve or so searches in which I was a participant-observer was unique. However, one search in particular—Northwestern University's—illustrates better than fiction the clash of formal machinery and partisan pressures in which American university presidents are made.

In simpler, less turbulent times, the presidential search process was handled pretty much the way an exclusive men's club chooses a new member. John G. Bowman, chancellor of the University of Pittsburgh in the twenties, recalls his own selection by this method in a chapter of his autobiography, entitled *You Must Go.* As Bowman writes:

> On an afternoon in October, 1920, in my work as the Director of the American College of Surgeons, I made a talk in Pittsburgh. The room was full of people, most of them interested in hospitals and the practice of medicine. After the talk, two or three men asked me to have dinner with them and some others that evening at the Duquesne Club. I said yes.
>
> In the evening I met about a score of men, most of them strangers to me, gathered around a table in a private dining room. During the dinner and after the dinner they asked questions: Do university presidents have business sense? What in your opinion is the top value of a college education? Is a college education a good thing for everybody? On and on the questions went. We had a lively talk and a good time. At about ten o'clock, one of the men at the table asked me to step into the hall with him. We went out of

the room and walked down the hall toward a window hung with heavy draperies.

We had covered only half the distance to the window, however, when another man of the group opened the door and asked us to come back. Then, all at the table again, George H. Clapp, head of Pittsburgh Testing Laboratories, said to me, "These men are trustees of the University of Pittsburgh. For some weeks we have been gathering information about you. We are glad now to invite you to become Chancellor of the University."

Bowman's story has a nostalgic quaintness to it, like a Dreiser novel, but the process he describes wasn't at all unusual. Douglas McGregor, president of Antioch College (1948–1954), once told me about his selection. He was in his office, at MIT, when his secretary told him that a Mr. Arthur Morgan was waiting to see him. McGregor knew little of Antioch and nothing of Morgan, who had been president of the college. Morgan offered him the presidency of Antioch after a few minutes of polite conversation which ran the gamut of the weather and the Charles River, snaking below the window of McGregor's office. They talked a while longer, and the next afternoon, the two of them took the train from the Back Bay Station for Yellow Springs, Ohio.

Even in the old days, naming one's successor was rather unusual (although after David Jordan left Indiana University to become president of Stanford, the Indiana board asked him to name not only his immediate successor but the next three presidents as well). The selection of a new president was most often the once-in-a-lifetime task of the board of trustees or the university corporation. Board members usually knew a few well-placed individuals to call on to suggest nominees, and people like Andrew S. White, the first president of Cornell University, Nicholas Murray Butler, or Chancellor William Tolley of Syracuse were frequently consulted. White, for example, personally picked two presidents for Michigan, one for Indiana, one for California, and one for Brown, and he also suggested the men who became the first presidents of Stanford and Johns Hopkins. In the 1920s, few boards reached their final decision without having consulted the Rockefeller Foundation. More recently, the Ford and Carnegie

Foundations have played crucial roles in identifying potential presidents.

But times have changed. Today, universities are expected to choose presidents in the open by a process which involves students and faculty in some meaningful way. The process is expected to be a democratic one. In my experience, what actually happens when universities choose presidents falls considerably short of that ideal.

Take Northwestern. Sometime in mid-November, 1969, my office received a call from the Chicago office of the management consulting firm of Booz, Allen & Hamilton. I didn't return the call until Thanksgiving week because of university pressures and more or less mindless lethargy. Also, a sense of low priority. I had mistakenly assumed that Booz, Allen was going to ask me to give a speech on management. I had stopped giving speeches on management sometime before, again because of university chores, and because I felt increasingly squeamish lecturing on problems I had written books about now that my own institution was suffering with symptoms very like those I was reputedly an expert in curing. Midspeech I tended to recall Auden's character "who lectured on navigation while the ship was going down."

When I finally returned the call, two days before Thanksgiving, I learned that the firm was contacting me as a consultant for Northwestern University. Their man wanted to know if I might be available for, "or at least interested enough to explore," the Northwestern presidency. "I'm not really sure," I answered. "I don't know a great deal about Northwestern and what it wants. But I would certainly be interested in discussing it with you." He asked me when I could come out to Chicago or make time available for him at Buffalo. I told him that it would be extremely tough to find any free time before mid-January, but that I would be in Cleveland over the Thanksgiving weekend, and if he could manage to come there, I could certainly see him on Friday or Saturday. He jumped at this and said that a man in their Cleveland office "who has been working on the Northwestern case" would be able to see me at my in-laws' home in Cleveland.

I cannot recall Booz, Allen's Cleveland man by name. He reminded me of many youngish management consultants I have known who worked for any of the Big Three: Booz, Allen & Hamilton; McKinsey & Co.; Cresap, McCormick and Paget. He was a Harvard Business School graduate, WASPish, attractive, crisp, alert, and formidably informed. We spent about three hours together and hit it off immediately. I felt he was as straightforward and honest as he could be about the Northwestern situation. He gave me an exhaustive survey of the present status and future of Northwestern, told me a little about the search activities to date, generally what kind of man they were looking for, how Booz, Allen came to be involved in "coordinating the search process," and an informal rundown of some of the people on the board of trustees. At the end of the discussion, I presented my "C.V." (my *curriculum vitae,* without which any academic man is *sans* identity). He told me that I would be hearing from them in the future.

What I learned at that time was this. Northwestern sounded first-rate. It had a healthy endowment, distinguished faculty, top-caliber students, and rich and prominent trustees. Moreover, it was untainted, so far, by "the troubles" its more distinguished neighbor, the University of Chicago, had experienced during the past three years or so. Undertaking the presidential search was an omnibus committee which consisted of nine trustees, four faculty members (elected by the faculty), and three students. Booz, Allen had collected the names of more than three hundred possible candidates from the faculty, the board of trustees, and students. The incumbent president, a former dean of the Northwestern Medical School, Rocky Miller, was close to sixty-eight years old, mandatory retirement age, and had been in office for about twenty years. He had been, according to my informant, a good, somewhat conservative president who was instrumental in bringing about a substantial building program at Northwestern. With Booz, Allen's help, the search committee had drawn up a three-page document describing the kind of man they were looking for to replace Miller. Dr. Right was married, between thirty-five and forty years of age, with a strong, broad academic background, administrative experience, vision, energy, good health, and an ability to talk with diverse constituencies, and someone who could keep the campus relatively free from disruption. Booz, Allen was in the act because the board felt that a good consulting firm, with a strong track record for executive "headhunting," could assist in the normally chaotic selection process. I also learned that Jim Allen, a Northwestern trustee, was the "Allen" in Booz, Allen & Hamilton.[1]

I was very taken with Northwestern after that talk and felt highly complimented by their interest in me. Of the colleges and universities that had contacted me in the past year, I had taken two of them seriously enough to visit, but they weren't in the same league as Northwestern as far as resources, talent, and future potential were concerned. Several factors contributed to my initial enthusiasm: Northwestern was in an important urban area which the university had virtually ignored in the past and with which it now wanted to get involved. The university had yet to achieve "greatness" in the same sense that Harvard or Yale or even Stanford had (Stanford most of all, since twenty years ago Stanford was in about the same league as Northwestern) despite their valid claim to top-rank faculty and students; there was not yet a truly unique character associated with Northwestern, no Northwestern*ness,* but the search committee obviously wanted such a character to emerge under new leadership. All of these suggested to me that Northwestern was on the verge of major growth, and that its new president would play a key role in directing its emergence as a major university.

Whatever doubts I had centered on whether Northwestern and I fit. From all I knew of Northwestern, it was conservative, rich, suburban, and Midwestern—what in the 1950s would have been called a "white shoe" campus. The trustees were the biggest question in my mind. For the most part they came from the large banks, law firms, and businesses in Chicago. They not only read the Chicago *Tribune* editorials daily, but according to my Booz, Allen & Hamilton man, they actually *believed* them. Boards have an understandable tendency to pick presidents who incline to agree with their values, if not their politics. (One study of 110 state college and university presidents showed that 88 percent stated their political affiliations as identical with the governors of their respective states.) Whether or not they would find me acceptable remained to be seen. That a consulting firm had been called in—no matter one of the best—also registered a blip of concern. Previous experience with presidential searches led me to believe that this was extremely atypical. So I had a few qualms, and I knew there would be surprises ahead. But even when one is prepared for surprises, they do surprise all the same.

A week or so later, Booz, Allen called again. Without really using these words, their man told me that I had passed the first hurdle and in a month or so I would hear from him again. Apparently, there were other people to screen, and if I stood up well, a contingent from Northwestern would visit me in Buffalo. He was extremely friendly, even solicitous. "Do you want any more information about Northwestern?" he asked before ringing off.

Sometime in the first half of January, 1970, I was called again. The search was going along nicely, my Booz, Allen contact reported. The list was now pared down to fifteen or twenty names. He said that I was still very much in the running, and, indeed, he wondered if I could meet with the chairman of the search committee, who was also a Northwestern trustee and chairman of the board of Harris Trust Company in Chicago; the associate dean of Northwestern's School of Speech and Drama, representing the faculty; and the president of the Northwestern Student Association. We set the time for February 4 in my office at Buffalo.

The group arrived in my office around 10 o'clock in the morning and left for the airport at 4 in the afternoon. Lunch was brought in. It was a six-hour talk show. After the initial awkwardness was smoothed over with orders for coffee or tea and talk of their flight, we settled down for some serious conversation. We covered a range of topics: from my attitudes about student participation, campus unrest, and my role at Buffalo, all the way to social activities and family. (They did not ask whether I had a "Hebrew strain," as one trustee of Penn State University had the year before when he visited me at Buffalo. At that time I had replied that not only did I have a strain, I was, indeed, a "Hebrew." "I knew it! I knew it!" the trustee had exclaimed enthusiastically.)

The time with the three search committee members was well spent. We covered a lot of ground, and it was clear that they not only permitted but encouraged me to indicate how I stand and who I am. They had read some of my books and articles, and referred several times to a revealing interview with me which had appeared in the February, 1969, *Psychology Today.* (Actually, candidates are rarely as satisfied with their interviews as I was. According to a recent study conducted by the Committee on Educational Leadership in New York State, only a small number of college presidents felt that the selection process had permitted them to show their strengths for the position in a significant manner. I figured that even if Northwestern were to favor someone else, at least they would do so with a knowledge of my strengths as well as my weaknesses.) My strategy in dealing with search committees, if it can be called a strategy, is to be myself as much as is humanly possible to be oneself with a group of caliper-eyed strangers. If they *really* want someone like me, then there is none better.

In any case, all of us seemed to enjoy the day. The faculty member was interested to learn if I cared more about academic pursuits than administrative affairs and whether or not I would work with, rather than against, the faculty. Faculty tend to view or want to view the president as their servant, not their seer, he said. The student clearly wanted a president who would take the students into account, not just in a token way, and he dwelt on problems of student involvement in the political processes of the university. We got into a little argument concerning the tenure of presidents. I said rather firmly that I believed in a term appointment of about seven years, as Kingman Brewster had recently advocated at Yale, rather than a lifelong appointment. This caused the chairman to ask if I really were interested in Northwestern and "might it not be a mistake to set a definite period of office?" The discussion was easy, flowing, informal, and without a great deal of anxiety.

The sticky points were all predictable. I asked them a battery of questions about Northwestern's culture and style, its financial status and outlook, what kind of guy "Rocky" Miller was, why they were interested in me (Why was I interested in them? they asked back, quite reasonably), what was Northwestern's relationship with other universities in the area and with the cities and state, how many "disadvantaged" students are admitted, about black studies programs in general, some notion of presidential discretion and power, the structure of the search committee, the weaknesses in the present organization of the university, an evaluation of the present administration, relations with alumni, and so on.

I *liked* them, particularly the student, who asked the most penetrating and direct questions. The faculty representative appeared to be gentle and perceptive and a remarkable listener. The chairman was the least relaxed and sometimes irrelevant, going off on a tack of his own which I couldn't always understand. He was clearly concerned about campus disorders, but he appeared to be a broad-minded and open man.

When the three returned to Northwestern, they met with Booz, Allen for a two- or three-hour "debriefing." On the basis of their reports, Booz, Allen prepared the following summary of their impressions:

Summary Comments from Subcommittee Meeting with Prospective Candidate

Warren G. Bennis
Acting Executive Vice President
and Vice President–Academic Development
State University of New York at Buffalo
Buffalo, New York

Academic Qualifications And Experience

Strong qualifications and credentials. Full professor—two institutions. Innovative and progressive. Some national eminence through extensive writing and lecturing. Favorably inclined to professional fields as well as academic—"cultural pluralism." Both a teacher and writer. Evidently strong faculty/student relations and sensitivity. Former President Meyerson's selection of him is tribute to his academic as well as administrative credentials. Buffalo and state university affiliation should not be held against him. The university has made many significant changes and improvements.

A dissenting opinion by one subcommittee member—"compulsive speaker and writer—perhaps has overdone it."

Executive Experience

Has handled key issues of the "number-two" job at Buffalo very well under heavy pressure. Clear understanding of how the administrative structure can complement, even aid, the development of new progressive currents in the field of higher education. Has played key role in the improvements made at Buffalo. Is an innovative administrator—many new programs evidently well planned and executed. Delegates well. Personally well organized; firm-minded; appears to be decisive and no impressions of limitations in leadership. Consensus opinion that he can contribute, but we must check his track record—Is he a senior *executive* or merely an able, persuasive administrator?

Magnitude of Present Management Responsibility

Experience is applicable but effectiveness has not been proven over extended period. Chief operating officer for past several months of sizable, complex, urban, public university. Larger in enrollment than Northwestern but not as broad in scope. Helped effect necessary and important changes from private to public institution management. Important to recall that he had significant experience as a professor and department chairman at a large, private university (MIT). Is not tied to any system of organization and expressed reservations on Buffalo multiarea provost system. Also raised questions on NU organization, but appears to be objective and open-minded. Works hard at managing and evidently is very demanding.

Personal Qualifications

Medium height, "tweedy," "modernish" but neat. Personal bearing "taller than his physical stature." Poised, articulate, charming, genuine, and businesslike warmth. A quality man. Smart, analytical—once he makes decisions will be demanding and perhaps unbending. Could work well at all levels. Will want authority but suspect he would be sensitive to others. Community-minded and gives impression that his wife is equally extroverted (although subcommittee did not meet her). While affable and thoughtful, there is clear indication of firm-mindedness and high opinion of self. [A possible, but not serious, reservation that his manner might not "sit well" with some faculty, trustees, and alumni.]

Apparent Interest in the Position

Very high. Really "did his homework" on NU. Asked extremely good questions. Sees Northwestern as logical and attractive next step for him. Sees the university as an academic institution of great accomplishment and potential. [Would want to come only if he were convinced he could ultimately become the chief executive.] Was sensitive to the importance of the relationship with the Chancellor—raised subject himself.

Summary Comments

A good meeting. Personally and professionally attractive. Handled himself well. Has a style that might be just right for the coming era at Northwestern. An impressive, innovative, competent, and accomplished administrator. May be limited in experience and a bit too aggressive and demanding, but is a strong candidate who should be considered further.

Similar profiles were drawn up on each of the stronger candidates and distributed among the entire search committee. (The members of the search committee had specially made black notebooks, with their names embossed in gold on the covers, in which to keep the profiles and all their other search materials together.) Of course, as a candidate I knew none of this at the time. What was happening at the Northwestern end of their yearlong search was told to me only recently by one of the on-campus participants.

Two weeks passed after my interview, and the Booz, Allen representative phoned again, this time with positive joy in his voice. The search committee wanted me to come to Northwestern *as soon as possible* to spend at least a day or two talking with people. He was annoyed by the fact that two key Northwestern faculty committees had "demanded" the right to interview all "serious" candidates, and he hoped I wouldn't mind if they were included in my visit. "There is now a short list of candidates," he told me, "and you are on the short list. Only the top four on the short list are being invited to the campus." I was leaving for Mexico City in a few days and told him that I would call him as soon as I got back. I called on February 23, and we settled on March 7 for my visit to Northwestern, the day of the eclipse.

Between my phone call to Booz, Allen on February 23 and March 7, the Buffalo campus exploded into one of the worst crises that have hit major campuses in America. Strangely enough, except for faculty member Edgar Z. Friedenberg's articles in the *New York Review of Books,* a squib in *Newsweek,* and an occasional story in the New York *Times,* the *Post,* and the *National Observer,* there was surprisingly little national coverage of the Buffalo crisis. Perhaps people were getting bored or inured to the guerrilla activities of students and the police (this was *before* Kent State and Jackson State), or maybe the Santa Barbara bank-burning about the same time eclipsed everything else. I don't know. In any case, the Buffalo campus experienced an unparalleled convulsion, with the local press reporting news of "fresh disasters" daily. More than 125 students, police, and others required medical attention. Forty-five faculty members were arrested and booked on three separate charges (the largest number of faculty ever arrested for a campus activity; the previous record was sixteen Harvard faculty arrested for protesting the Spanish-American War). Another six faculty members were arrested on other counts. More than fifty students were charged on criminal counts. Over $300,000 in property damage was reported.

At seven that morning, March 7, on the plane for Chicago, I

wasn't at all certain that I should be going. I had averaged only two hours sleep per night over the last three weeks. (If there is any advice I would give campus administrators during a crisis, it is this: GET SLEEP. The kids are younger and there are more of them. They can run in relays.) I felt like limp pasta as I flopped into my seat. I was no more up to what I knew would be a grueling day of scrutiny that I was to another day of manning the barricades on my own campus.

Worse by far than the fatigue was the recognition that I *should* remain in Buffalo that day in order to stave off a decision made by certain members of the university's administration the night before. The acting president, Peter Regan, and some of his advisers were convinced that the only way to stop the student disruption was through a massive police intervention, virtually an occupation of the campus, by four hundred Buffalo city police. I was convinced this would be a catastrophic mistake, that it would destroy whatever legitimacy and trust the present administration tenuously held; that it would depress morale below the tolerable limits; and, finally, that it would be playing into the hands of the most militant students, who badly needed another clumsy overreaction by the administration to survive. Calling in four hundred police was the single way to "radicalize" the majority of moderate students, forcing them to join and augment the usually thin ranks of committed revolutionaries; at Buffalo, a corps of perhaps 100 students out of a total student body of almost 24,000.

I had argued vehemently against the police intervention, for these and other reasons, for the preceding eighteen hours without success. When I left Buffalo the police were expected to move onto the campus twenty-four hours later, Sunday morning, March 8. My only hope—or rationalization—in leaving town was that the acting president would recover from this lapse of judgment when he could get some sleep, and when he was no longer involved in a "win-lose" argument with me, his number-two man.

All these things were on my mind when I finally met my Booz, Allen contact at O'Hare Airport, at 7:30 A.M., Chicago time. He escorted me directly into the chauffeured Cadillac owned by the search committee chairman and drove to the chairman's house for coffee. On the way, I was shown the schedule for the day. It was more clogged than usual because I had insisted (and everyone fully understood) that conditions at Buffalo were such that I could spend only one day for this visit. After coffee, I was to be driven over to the president's house for a two-hour talk with Rocky Miller. Then the chairman would pick me up just before noon and drive me out to the Glenview Country Club, where I would have lunch with a group of about ten trustees. (Booz, Allen proposed a later meeting with several absent trustees in Palm Beach.) Following lunch, I would meet with the two insistent faculty committees who "demanded" the opportunity to interview and register their responses to all "serious" candidates. Following that, I would have dinner with all those faculty members and students on the search committee who had not been able to come to Buffalo in February. This series of meetings would last through 9:00 P.M., allowing just time enough to return to O'Hare for the 10:05 P.M. flight back to Buffalo.

The man from Booz, Allen was visibly glad to see me. I remember him rubbing his hands together in delight when he told me that the search was down to four candidates, "two insiders and two outsiders." This was more information than I had expected, though I had heard that Northwestern's dean of the arts and sciences, Dr. Robert Strotz, was a very strong contender. He also surprised me when he mentioned the name of the other inside candidate, a man whom *I* was trying to recruit for a top administrative post at Buffalo. When I said that we, too, were interested in him, my Booz, Allen contact seemed both surprised and annoyed. "Haven't you guys looked seriously into his record here? Haven't you called *anybody* from Northwestern about him?" I replied that I wasn't on the search committee but I had every reason to believe that they had taken all the necessary steps, including informal "prowling," in order to arrive at their assessment. Our mutual candidate, it turned out, was the man responsible for the Northwestern "faculty intrusion" into the university's search for a new president. He was generally making life miserable for the search committee, which did not, of course, endear him to

Booz, Allen. My internal radar, dormant throughout the lulling airplane trip, was suddenly reactivated by the inappropriateness of the Booz, Allen man's remarks. As he grumbled on about faculty troublemakers, the candidate began to sound better and better to me. (He eventually accepted the position at Buffalo, by the way.)

The day was predictably grueling. The two hours with Rocky Miller were extremely cordial but basically a "nonevent." We covered nothing of consequence, although I tried on several occasions to bring him out. We spent most of our time in his car, driving around the campus. He pointed out buildings, told me what year they went up and which of the trustees had paid for them. I hadn't realized how bland it all was until Eva Jefferson, one of the students at dinner (soon to be president of Northwestern's student body), asked me what I thought of him. I really didn't have any response at all, except that he had certainly helped raise a lot of money for buildings.

The two faculty meetings were condensed into one session because lunch took longer than was expected and because the search committee chairman had trouble finding the room in which I was scheduled to meet with the two "ad hoc" faculty committees. So instead of meeting with them back to back, I saw all twenty-five of the faculty at the same time. That was a good session, as I remember. The questions were sharp and incisive, and I was heartened by an intuitive feeling that they were looking for someone like me. The dinner with the four faculty and three students was also active, penetrating, and pleasant. In addition to the drama and speech dean, who left early, the other faculty members included Raymond Mack, a first-rate sociologist and head of the Northwestern Urban Center, and, as I remember, faculty members from the Technology Institute and the health science areas. I felt very keen support from the students, moderate to strong support from the faculty.

Lunch at the country club provided the one remarkable moment of the day. It was a beautiful early spring day, with the eclipse throwing the bare-limbed trees into relief in the strong but muted sun, as if the sun were filtered through the edge of a fingernail. At times, I could barely look through the windows because of the strange brightness of the light. About ten people were present, including the omnipresent Booz, Allen man, the search committee chairman, and a representative of the Northwestern alumni organization. He was memorable because he was clearly the youngest in the group. The rest, with one exception, were in their fifties and over. But they were a handsome bunch of people, extremely cordial for the most part.

After Bloody Marys and sherry, we sat down at the table. Just as I began attacking the fresh fruit cup, the person seated two seats to my left, whom I remember as the alumni representative, cleared his throat and floated a question down to me. I was tempted to let it pass, but it was evident that he had been working on that question a long time, and, I thought—mistakenly, in retrospect—that taking a cut at it would get him over the embarrassment the question seemed to cause him.

"Who," he asked, "do you think are the three greatest university presidents and why?"

I returned my spoon with a melon ball resting on it to the plate and said, looking up into the oak beams. "Well, Howard Johnson of MIT, for one. In my view he is the administrator-manager *par excellence.* And imagine overcoming a name like that."

That seemed to relax everybody, and I continued. "My second choice would have to be Kingman Brew . . ." At that very point, somewhere between the "Brew" and the unuttered "ster," the man opposite me began to choke as if something were caught in his throat. Two red-jacketed waiters ran over to him and started pounding him on the back. This lasted a good thirty seconds, until he seemed to recover his breath. His breath but not his composure. As he came up into a vertical position for air, the man shouted something to me. I couldn't hear him, although he wasn't more than three and a half feet across the table from me, and as I leaned forward, the ball of honeydew that had lodged in his throat at the mention of Brewster left it at a muzzle velocity of at least 1000 feet per second and smashed against my forehead.

At the moment of impact, the Booz, Allen representative, seated directly to my right, kicked my leg, and I began to wonder if this was some kind of perverse stress test they gave to all candidates. As the waiter dried my face nervously, my red-faced assailant increased the volume. He screamed, choking again. "DID YOU SAY BREWSTER? WOULD *YOU* KEEP THAT IDIOT COFFIN ON YOUR PAYROLL?"

I ducked involuntarily and then replied, more dogmatically than I actually felt about the whole business, that, yes, the Reverend Coffin apparently serves an important purpose at Yale, despite his radical views, or at least, Brewster thinks so, and furthermore, I'm not sure that Brewster has the power or right to countervene on what are basically faculty prerogatives, and . . ." My questioner's coughing had subsided to heavy breathing, but his face was still alarmingly red. A trustee seated at the head of the table asked in a commanding voice if I wouldn't turn to my third choice, and we continued, almost pleasantly, as if what had happened were a trifling *faux pas* that we had collectively agreed to ignore.

The "honeydew statement" aside, the Northwestern search to this point is more or less typical of how most universities go about selecting a new president. The only unusual feature, as I said earlier, was the use of an outside consulting firm to coordinate the search process. (Later on this backfired.) Of the 2500 or so accredited colleges in the United States, only the most parochial (say, Bob Jones University in South Carolina) would proceed on a presidential search without a faculty, student, and possibly an alumni committee, working with a small group of trustees.[2] Northwestern also had alumni representation on the board. When a community college is searching for a president, the committee almost always includes prominent members of the community.

The search that Harvard undertook before selecting Derek Bok to succeed Nathan Pusey may well emerge as the new model for selection processes, at least at major institutions. Harvard's Committee on Governance (appointed by President Pusey in September, 1969, a product of the spring crisis) circulated a fifteen-page booklet, *Discussion Memorandum Concerning the*

Choice of a New President. Published in April, 1970, it called for the most thorough participation in the presidential search known to Harvard, certainly, and perhaps known to any campus. Aside from an incredible number of consultations with "key groups," both inside and outside the University, the Harvard Corporation started its search with the distribution of some 200,000 letters inviting suggestions for candidates from (among others) faculty members, students, key alumni, and employees. This correspondence alone took almost the full time of a professional staff member with a large staff of assistants and clerks. That first step was only a small part of the total effort, which cost an estimated half million dollars. (By way of comparison, James Conant was selected after only one appearance before the total search committee, during which he gave a clear analysis of the man needed and urged the candidacy of one of his closest friends.)

In one magnificently prescient paragraph entitled, "How might the foreseeable negative consequences be minimized?" the Harvard *Discussion Memorandum* lists these five stumbling blocks which may cripple a search committee:

1. Sheer volume of work.

2. The selection process may divide and polarize rather than unify the university.

3. Candidates surviving the scrutiny of many diverse groups may be the same hackneyed names who always turn out to be unavailable—what I refer to as the "John Gardner" syndrome—or, among those who are available, they may be essentially "low risk," mediocre candidates.

4. A "credibility gap" may occur between the search committee and various groups within and without the university over the extent to which their advice is being sincerely sought, objectively evaluated, and imaginatively interpreted.

5. Potential candidates may be alienated by premature publicity, gossip about their candidacy, and vigorous opposition, even if ill-informed and limited.

Northwestern's search process left none of these "negative consequences" unturned. In fact, it uncovered several "negative consequences" undreamt of in Harvard's list.

Quite often, especially if the university is prestigious, there is a good deal of publicity that attends and sometimes complicates the search. In the Northwestern case, for example, the student newspaper, the *Daily Northwestern,* managed, somehow, to obtain dossiers on each of the three finalists as well as their ranking. How they succeeded is one of the miracles of the age, since all search committee members are sworn to secrecy and the candidates themselves are usually in no mood to discuss the race. For obvious reasons. Nobody wants to have "honorable mention," even on the "short list." Active candidates try to appear majestically aloof from the politics of candidacy. Overt campaigning is as alien to the academic man, and as endemic, as it is to the College of Cardinals during papal election. All that is missing when a university picks a president is the puff of smoke.[3]

The *Daily Northwestern* scoop of April 15 revealing the names of the three finalists—New York University's Chancellor Allan Cartter, Northwestern's Dean Robert Strotz, and myself—was immediately picked up by all the Chicago dailies and in the hometown newspapers of the two outside candidates. The next day the student paper ran an editorial scoring the search committee for ignoring student opinion in the selection process so far. Everyone concerned refused to comment, although the leaked story was completely accurate, so accurate, as a matter of fact, that one candidate decided to withdraw from the race shortly after the story appeared.

Owing to poor weather conditions, the 10:05 plane from Chicago to Buffalo on March 7 was unable to land, and I arrived in Buffalo around 11 A.M., Sunday morning, March 8. The acting president had called a noon meeting of all faculty, administrative officers, and student officers, and I rushed over to the campus to learn what was going on. The entire perimeter of the campus was surrounded, bumper to bumper, with Buffalo City Police vehicles. As I walked into the building where the meeting was to be held, policemen were already marching across the campus, Army-style, twelve to sixteen per group in columns of two. When I arrived, Acting President Regan was explaining his reasons for calling the police onto the campus, a speech greeted for the most part with a lobotomized silence. Only three people spoke up. The vice president for student affairs said that "it was about time," and a member of the faculty senate who supported Dr. Regan's action reported that "my department would certainly support the police action." Mark Huddleston, president of the Student Association, said that he and the other students were totally against the police occupation, that he was even more thoroughly disgusted with the acting president for going back on his word that students would be consulted on all decisions related to police intervention. The meeting dissolved after Huddleston's statement.

Afterwards, I walked with Dr. Regan to a nearby office and told him that I intended to resign, not only to disassociate myself from the police intervention, but for other reasons that made my future cooperation with his administration impossible.

In writing my letter of resignation that evening (which was personally delivered to Regan the next morning, March 9), I was not totally unaware of the consequences of the act and how it would be perceived, both inside and outside the university. If there were still any doubts about the Northwestern presidency, I felt this act could be interpreted, particularly by the Northwestern trustees, in any number of unfavorable ways, from being "too permissive" or "soft on law and order" to "desertion." My fears were confirmed during the next few months in a series of unusual exchanges with Northwestern.

On Wednesday, March 11 (the day after the story of my resignation appeared in the New York *Times*), Booz, Allen called to say that the tentative meeting called in Palm Beach, Florida, during which I was supposed to meet a contingent of Northwestern trustees who spend the winter there, had been canceled. Their man, no longer a disembodied voice, also pumped me about the resignation, clearly trying to establish the background for my decision. Before the conversation ended, he assured me that he would "keep in touch." Around the end of the month, having heard nothing in the meantime, I phoned him to find out how things were progressing with the search. He seemed very flustered and said that things were in a mess and that he didn't have any news for me as yet. He said that he wished he'd never gotten involved in this mess and that he hoped to have a chat with me, "man to man, after this whole thing blows over." "Then," he said, "I can really level with you." I thanked him and hung up, wondering what was really going on.

So, apparently, did the *Daily Northwestern.* An editorial in late April criticized the board for retaining, presumably at substantial cost, a management-consulting firm whose senior partner, James Allen, also happened to sit on both the Northwestern board and the search committee.

Between late March and the first week in May, the only news I received about the Northwestern candidacy came through informal sources. A student journalist at Northwestern called to extract a story from me and in the process related a good deal of inside information. On the basis of a purloined copy of the complete search-committee proceedings and other leaks, she indicated that the students and most of the faculty favored me, but that the board of trustees, ultimately responsible for naming the president, was, as expected, polarized; that Robert Strotz, who has been at Northwestern for practically his entire academic life, was seen by the board as the "safest" candidate; that Cancellor Cartter of NYU was moderately acceptable to the board and some students, but that the faculty was opposed to him. (I heard in early May, from Booz, Allen, that Cartter withdrew after the publicity upon the advice of his wife.) The students were so upset about the prospect of Strotz (for reasons which were unclear to me) that they intended to do everything possible to block his appointment and to insist that only Cartter or I was acceptable. They intended to use all means necessary to delay, obstruct, and ultimately subvert the Strotz appointment. According to the girl, the students were not ecstatic about Cartter, but were sure that the faculty would blackball him. I was, she thought, the candidate most likely.

Leaks often ended up as headlines in the Buffalo press and on local TV. Both of the Buffalo dailies used their Chicago contacts to ferret out any news from there, and the Chicago *Tribune* sent a reporter to Buffalo to do a piece on me for background. On May 3, the *Trib* ran side-by-side stories on both Cartter and me under the headline "TWO EDUCATORS SEE OTHER MAJOR ROLES FOR U.S. IN-STITUTIONS." Excerpts appeared in the Buffalo papers.

I also heard, about this time, that the Northwestern trustees were in touch with some of the Buffalo Council members, establishment types for the most part, and that the chairman of the Buffalo Council was saying publicly that he had inside information and "knew that Bennis was not going to be appointed president of Northwestern." The precise nature and frequency of communication between the two groups were not known to me. I did hear that there was a good deal, none of it particularly helpful to my Northwestern chances. Meanwhile, a strong contingent on the Buffalo campus began urging me to run at home against odds far greater than Northwestern held for me.

Sometime around mid-May, I decided to call the chairman of the search committee directly, rather than going the Booz, Allen route. It was not particularly pleasant to initiate these calls. I always felt that it made me appear more eager than I really felt. On this occasion I called because two other universities were interested in "exploring" presidencies with me, and I wanted to be certain that Northwestern was really out of the question before investing time and energy on other prospects.

The committee chairman was reported by his secretary to be "tied up," and did not return my call. Booz, Allen's man did instead. If his mood in late March was disconsolate, he was practically teary on this occasion. He startled me by suggesting that I call the committee chairman. "Tell him that you want to withdraw," he advised after I remarked that I was getting concerned about the search's lack of progress, information, and the other spooky vicissitudes of the search process. I asked him what I would learn or gain by withdrawing my name at this point. "Well," he said, "this way they'll know you're really serious." "Is there any doubt about my *seriousness* at this point?" I asked. "Well, you never can tell. And it will serve them right."

My level of paranoia, usually abnormally low, was rising, along with increasing doubts about Northwestern. I did call the committee chairman back a few days later. He began by apologizing for not returning my call. Trying to get clear and direct information from him in person was difficult enough, but on the phone it was like trying to nail a chiffon pie to a wall. I did not mention my most recent conversation with Booz, Allen, nor did he. He finally said that the search was taking longer than they had expected, but that I was still in the running. "Hang in there," he encouraged me.

In early July, I called Booz, Allen for the last time. Their man indicated that he really didn't know what was going on at this point, but that I should consider myself "out of the running." The following day I received a call from the chairman of the search committee, who, to my amazement, informed me that I was still "a very active candidate" and that he hoped to have word for me no later than July 21 or 22. He also reported that he had just been elected president of the Northwestern Board of Trustees. In order to facilitate the whole search process and to select a president no later than the third week in July, when the full board (about forty members) had its regular meeting, he was going to meet individually with each trustee. I told him that I was leaving for Europe on the sixth of August for a month, but he assured me that I would hear from him no later than July 21.

I should have listened to Booz, Allen.

In the week preceding July 21, Buffalo radio and TV stations reported that Northwestern would appoint a new president at its regular meeting of the board on July 20 and that the race was between Dean Strotz and myself. On Saturday morning, July 18, Eppie Lederer (better known to millions of her readers as Ann Landers), an old friend, called to tell me of several articles which had appeared in the Chicago press on July 17.

Eppie was convinced that student leader Eva Jefferson's strongly worded statement against Strotz as the "old guard" choice would discourage the Northwestern board from appointing Strotz. In her view, I still had a pretty good chance for the post, and she advised me not to "give up hope." I told her I thought that for once she was dead-wrong, that my chances for the Northwestern presidency were nil, particularly since the Kent State–Cambodia crisis. Besides, my own university had, within the past three weeks, appointed a president over even stronger student *and* faculty objections. As it turned out, Ann Landers *was* wrong, a first perhaps, although I didn't know this until the first week in September.

When my family and I sailed for Europe on August 6, I was still in the dark about the Northwestern selection. The committee chairman's promise of "getting in touch no later than July 22" never materialized. After the front-page hullabaloo over student opposition to Strotz, I wagered that the whole search process to date would be scrapped and revived only with the appointment of a new search committee.

In early September, I returned from Europe via Miami to give a talk to the American Psychological Association. At a convention cocktail party, a vaguely familiar-looking man came over to me and reminded me that he was a member of one of the faculty groups at Northwestern who had interviewed me. "I'm just so sorry that you turned us down," he said. I wasn't sure how to respond to that gracious and tactful opener. So I said, "I appreciate your tact and graciousness, but I was never asked. Was it Strotz?" He said that it was.

To this day, I have not received official word from Northwestern concerning my candidacy or Strotz's appointment. Not from the search committee or from Booz, Allen. Finally, out of sheer perverseness, I suspect, I sent the following letter to the committee chairman:

> When I returned from Europe on the 7th of September I quite by accident heard that Dr. Strotz was appointed President of Northwestern University. For any number of reasons, I am certain that you are delighted that the search is over and that you have found a first-rate person to lead Northwestern at this point. I congratulate you and the Board on its choice and wish you the very best of luck—which all of us will be needing—in the years ahead.
>
> In light of my own candidacy for the post, I wonder if you could write me a note, totally off the record, that would help me understand the reasons for the Board's final selection. Obviously, I am aware that privileged communication and tact, and the usual practices, do not allow the frankest or most open discussion of this

issue or the reasons why any candidate is chosen over any other, but I would like and would deeply appreciate "feedback" from you especially.

I might also say that I was surprised to hear the news in the manner I did. You had originally told me that you would be back in touch with me before leaving the country and when I did not hear from you through August 6 when I left with my family for Europe I had just assumed that no decision would be made until I had been told, before learning about it in the inadvertent way I did.

Anyway, I would appreciate hearing from you whenever time permits.

Sincerely yours,
WARREN G. BENNIS

He has never answered my letter.

Points to Remember on Choosing a College President

1. *Remember that there is no single quality, trait, characteristic, style, or person that guarantees presidential capability.* A century and a half of psychological research confirms this point. An Ivy League degree or a "low profile" is not in itself going to ensure the bearer of success in dealing with an adamant board or angry students. Being from the "outside" is no talisman either. The outsider may fail if he is unable to master quickly the special terrain of his institution—fail just as dismally as the insider whose judgment is skewed by partisan loyalties held over from his pre-presidential days. There is no one presidential "type," no presidential personality. The time is past when a Stanford or a Columbia can be described as the lengthened shadow of any one man.

Many different approaches to university management have been successful in the recent past. Among possible presidential styles are:

The problem-solver/manager. Howard Johnson, the retiring president of MIT, has used this approach most successfully. Johnson's concern has been, How can I identify problems (real problems, not temporal issues) and engage the best minds and most important constituencies to work on them?

The managerial style is often confused with that of *low-profile/technocrat.* Similarities are superficial. Instead of putting the right people to work on the right problems, the technocrat tries to find *systems* that will somehow transcend human error. The concerns of the technocrat are all pragmatic. He cuts through moral and ideological dilemmas with a callousness that soon has students and faculty aligned against him.

The leader/mediator. Based on the labor relations model, this style is just coming into its own. If one conceives of the university in terms of constituencies seeking to maximize self-interest, a place where there is no way to make decisions without pleasing some and making some angry, then this style is very effective. In fact, there may be no decent alternative. A number of men from industrial relations backgrounds have become very successful presidents recently, most notably, Robben Fleming at the University of Michigan. (Howard Johnson also has a labor relations background; he first came to MIT as dean of the School of Industrial Management.) A problem that such men have is that since they cannot help making one side on any issue angry, at a certain point in time the accumulated anger overtakes the goodwill. So the tenure of such a president will be problematical unless he possesses, in addition to mediating skills, a degree of charisma that keeps him personally above conflict.

The value of labor relations experience does not seem to have escaped the Harvard Corporation, which recently named Derek C. Bok as Harvard's twenty-fifth president. Bok, who has been dean of Harvard Law School, is an authority on labor law and has been an arbitrator in several major disputes.

The collegiate manager. This is the style of the academic administrator in the strict sense of the term—the man whose primary commitment is to a scholarly discipline, who assumes the presidency as a faculty colleague rather than as a professional administrator. This man is very like a *representative* leader. The model is Parliament, with the faculty as the House of Commons and the trustees rather like the House of Lords.

Faculties have already acquired substantial influence at the great American universities. Nonacademic leaders can forget just how powerful the faculty is within these institutions. General Eisenhower, during his Columbia presidency, had to be reminded by the vice-chairman of the faculty senate that "the faculty *is* the university, sir!"

The communal-tribal or postmodern leader. Leaders of this style are emerging in many of our institutions, not just universities. The academic leader of this style usually heads a college, not a university. The tribal leader typically identifies strongly with students; he not only backs them, he often joins with them, whether on marches to

Washington or on strike. He is himself an activist, and likely to be young. John R. Coleman of Haverford or Harris Wofford at Old Westbury College (he is now at Byrn Mawr, where his style may be somewhat different) are examples of this style.

The charismatic leader. John Summerskill, who preceded Robert Smith (who preceded S. I. Hayakawa) at San Francisco State, was a charismatic president, but the exemplar of this style is Kingman Brewster of Yale. Brewster's personal attractiveness makes it possible to transcend obstacles.

In addition to these more or less acceptable presidential styles, there are several other possible approaches to university governance that should be mentioned. The following styles are currently out of favor or actually undesirable, but all of us have known men who practiced them.

The law-and-order president. Hayakawa, with his tam o'shanter and megaphone, is the epitome of this style. Ronald Reagan's behavior as the self-selected head of the University of California is also fairly typical.

The absentee-pluralist. This style, rapidly losing favor, has been highly regarded in the past. The president who adopts this approach sees his primary function as raising money for buildings and other needs and appointing competent subordinates. He hires those he considers to be good deans, spends his time on ceremonial functions, and "lets things happen." This is a spectacularly effective model when the university is rich, the subalterns capable, and the students and faculty relatively homogeneous and docile. In other words, if the university is like an elite men's club or the year is 1915.

The bureaucrat/entrepreneur. This style drives faculty to despair. The academic entrepreneur *par excellence* was Millard George Roberts, who with phenomenal *chutzpah* transformed a marginal sectarian college in the Midwest into a booming financial success and a national scandal. Before the bubble burst, Roberts succeeded in running Parsons College in Fairfield, Iowa, less like an academic institution than like a railroad. A *Swiss* railroad.

When all else fails, and the search committee and board cannot reach agreement on any of the above presidential styles, there is always the *interregnum* (or Pope John) solution. Interregnum leaders often do much better than might be expected. A good secular example is Dr. Andrew Cordier, who surprised almost everybody with his able management at Columbia.

There is at least one other presidential style, that of the *Renaissance or protean man.* This is the elusive superman that so many search committees pursue, the man who is all things to all constituencies. The protean president can role-play, presenting himself as a *communal-tribal leader* on some matters, a *bureaucratic-entrepreneur* on other matters, and on still others a *problem-solver/manager.* One of these protean men can also make life excruciatingly difficult for his constituents, who never know from one day to the next exactly what to expect.

2. *Determine the university's particular metaphor, the collectively held image of what the university is or could become.* Just as there are a number of successful presidential types, there are many university metaphors. The State University of New York at Buffalo comes close, in my view, to a "labor relations" metaphor. There are many other usable metaphors: Clark Kerr's "City," Mark Hopkins' "student and teacher on opposite ends of a log," "General Systems Analysis," "Therapeutic Community," "Scientific Management," my own "temporary systems," and so on, competing with the pure form of bureaucracy.

3. *Forgo the costly hit-or-miss search and tailor the search process to the special requirements of the individual university.* Once the university's metaphor—its collective self-image or ideal self—is determined, the type of president sought is automatically less problematical.

The university's metaphor should determine not only the style of the president sought but also the composition and relative weighting of the search committee. For example, if the university requires a collegiate manager, an individual with strong academic qualifications and faculty identification, then faculty should have the decisive voice on the search committee.

As corollary to Number 3, it is increasingly clear that *a presidential search committee should undertake only an intelligently limited canvass, not a national quest.* When a university picked a new president every twenty years or so, it was reasonable to underwrite a far-flung search, sparing no effort or expense to screen conceivable candidates. But the national search is beginning to appear as extravagant as the elaborate inauguration.

4. *Assuming that the search committee is representative, the committee should select the president as well as screen candidates.* Demoralizing conflicts can be avoided by making sure that trustees serving on the search committee are powerful enough and numerous enough to represent the total board throughout the search. This seems to be the only sure way to avoid the enormous frustration that results when a board of trustees overrules the decision of a responsible and representative search committee.

Notes

[1]Booz, Allen had actually created the presidential vacancy, it seems. It was Booz, Allen that recommended that Miller be moved up from president to chancellor as part of an administrative reorganization plan commissioned earlier by the Northwestern board.

[2]A survey reveals that in 1939, faculty were consulted in the selection of 29 percent of the college presidents then in office. By 1955, that figure had risen to 47 percent. By 1965, faculty were formally represented by a committee to advise the board in 65 percent of the cases.

[3]Confidentiality was actually poor, I've learned since. A member of the search committee told me that he was amazed at the number of Northwestern faculty who could recount the details of any meeting of the committee. Alumni talked even more openly. An alumnus broke the story of a three-way race to the *Daily Northwestern*.

10
When to Resign

Friends from the South talk about having to make "hurtin' decisions," those choices which tear at the soul but must be made. None is more painful than deciding you must leave a once beloved institution, whether it's a marriage or a job. This piece explores the high cost of quitting and the higher cost of staying on in a role in which you are no longer effective. The essay was written with Patricia Ward Biederman.

No matter how often Daniel Ellsberg reminds the public that not he but a seemingly endless war in Indochina is at issue, I find that it is Ellsberg the man who touches the imagination. One can't help speculating on his personal odyssey from loyal insider to defiant outsider, from organization man to prison-risking dissident. It is the process of that change of heart that fascinates me. What interaction of man and organization produces a commitment like the younger Ellsberg's and then leads only a few years later to equally passionate rejection? How much, I wonder, of the Ellsberg affair is idiosyncratic and how much reflects general principles of organizational life? After all, Ellsberg is not the first government adviser to become suspicious of the work in which he has engaged. What is singular about Ellsberg is that he has found a dramatic way to make his dissent articulate. The organizational ethic is typically so strong that even the individual who dissents and opts for the outside by resigning or otherwise dissociating himself does so with organization-serving discretion. Ellsberg may not have broken the law, but he surely did something more daring. He broke the code. He has not only spoken out, he has produced documentation of his disillusionment.

The stakes are rarely as great, but many people who work in

large, bureaucratic organizations find themselves in a position similar to Ellsberg's. They oppose some policy, and they quickly learn that bureaucracies do not tolerate dissent. What then? They have several options. They can capitulate. Or they can remain within the group and try to win the majority over to their own position, enduring the frustration and ambiguity that goes with this option. Or they can resign. Remaining can be an excruciating experience of public loyalty and private doubt. But what of resigning? Superficially resignation seems an easy out, but it also has its dark and conflictful side. As the Stevenson character in *MacBird!* says:

> In speaking out one loses influence.
> The chance for change by pleas and prayer is gone.
> The chance to modify the devil's deeds
> As critic from within is still my hope.
> To quit the club! Be outside looking in!
> This outsideness, this unfamiliar land,
> From which few travelers ever get back in . . .
> I fear to break; I'll work within for change.

If resignation is the choice, the problem of how to leave, silently or openly voicing one's position, still remains.

These options are a universal feature of organizational life and yet virtually nothing has been written on the dynamics of dissent in organizations, although a recent book by Harvard political economist Albert O. Hirschman almost single-handedly makes up for past deficiencies. Oddly enough, the book still remains "underground," largely unread by the wide audience touched by the processes Hirschman describes. I first began seriously considering the question of resignation and other expressions of dissent as organizational phenomena in the Spring of 1970. At that time I had just resigned as Acting Executive Vice-President of State University of New York at Buffalo. As so often happens, my interest in the phenomenon grew out of unpleasant personal experience. I had resigned in protest against what I considered undue use of force on the part of the University's Acting President in dealing with a series of student strikes on our campus that spring. In my case, resigning turned out to be a remarkably ineffective form of protest.[1] When I tried to analyze why, I found that my experience was hardly unique, that most large organizations, including government agencies and universities, have well-oiled adaptive mechanisms for neutralizing dissent. The individual who can force the organization into a public confrontation, as Ellsberg did, is rare indeed.

The garden-variety resignation is an innocuous act, no matter how righteously indignant the individual who tenders it. The act is made innocuous by a set of organization-serving conventions that few resignees are able (or even willing, for a variety of personal reasons) to break. When the properly socialized dissenter resigns, he tiptoes out. A news release is sent to the media on the letterhead of the departing one's superior. "I today accepted with regret the resignation of Mister/Doctor Y," it reads. The *pro forma* statement rings pure tin in the discerning ear, but this is the accepted ritual nonetheless. One retreats under a canopy of smiles, with verbal bouquets and exchanges, however insincere, of mutual respect. The last official duty of the departing one is to keep his mouth shut. The rules of play require that the last word goes to those who remain inside. The purpose served by this convention is a purely institutional one. Announcement of a resignation is usually a sign of disharmony and possibly real trouble within an organization. But without candid follow-up by the individual making the sign, it is an empty gesture. The organization reasons, usually correctly, that the muffled troublemaker will soon be forgotten. With the irritant gone, the organization pursues its chosen course, subject only to the casual and untrained scrutiny of the general public.

The striving of organizations for harmony is less a conscious program than a consequence of the structure of large organizations. Cohesiveness in such organizations results from a commonly held set of values, beliefs, norms, and attitudes. In other words, an organization is also an appreciative system in which those who do not share the common set, the common point of view, are by definition deviant, marginal, outsiders.

Ironically, this pervasive emphasis on harmony does not serve organizations particularly well. Unanimity leads rather quickly to stagnation which, in turn, invites change by nonevolutionary means. The fact that the organizational deviant, the individual who "sees" things differently, may be the institution's vital and only link with, for lack of a better term, some new, more apt paradigm does not make the organization value him any more. Most organizations would rather risk obsolescence than make room for the nonconformists in their midst. This is most true when such intolerance is most suicidal, that is, when the issues involved are of major importance (or when important people have taken a very strong or a personal position). On matters such as whether to name a new product "Corvair" or "Edsel," or whether to establish a franchise in Peoria or Oshkosh, dissent is reasonably well tolerated, even welcomed, as a way of insuring that the best of all possible alternatives is finally implemented. But when it comes to war or peace, life or death, growth or organizational stagnation, fighting or withdrawing, reform or status quo—desperately important matters—dissent is typically seen as fearful. Exactly at that point in time when it is most necessary to consider the possible consequences of a wide range of alternatives, public show of consensus becomes an absolute value to be defended no matter what the human cost.

Unanimity, or at least its public show, is so valued within the organizational context that it often carries more weight with an individual than his own conscience. Thus, we see in the March 31, 1971, issue of The New York *Times* that "Muskie regrets silence on war" and wishes that he had made public as far back as 1965 his "real doubts about involvement in the Vietnam war." Instead, he said, "he voiced his concerns privately to President Johnson." "There are two ways," he said, "and they're both legitimate ways of trying to influence public policy. And I can guess the tendency is, when the President is a member of your own party and you're a Senator, to try to express your doubts directly to him, in order to give him a chance to get the benefit of your views." Senator Muskie said he often had done that, "but wished that I'd expressed my doubts publicly at that time." The article goes on to say that Muskie "was far less hesitant to criticize President Nixon's conduct of the war." In an adjoining article about Humphrey, The *Times* reports him as describing to a student audience "publicly for the first time the pressure he had been under from President Johnson not to speak out on the Vietnam issue. Many times during the first month of the 1968 campaign, he recalled, he had wanted to speak out more forcefully on the Vietnam issue only to be dissuaded by the President. This, he said, posed a personal dilemma. On the one side, he said, he saw his chances for winning the Presidency slipping away. But if he sought headlines on the Vietnam issue by taking a more critical stance, he said, he was being warned by the President that he would jeopardize the delicate negotiations then under way to bring South Vietnam and the Vietcong to the Paris negotiating table."

"That's the God's truth. . . . How would you like to be in that jam?" Humphrey asked a student.

Actually, Humphrey's "jam" is a classic one. A member in good standing of an organization, in this case the Johnson Administration, suddenly finds himself opposed to his superior and his colleagues in regard to some policy. If the policy is relatively unimportant or not yet firm, the objection may be absorbed by bargaining or compromise. If the issue at stake is actually trivial, it may simply be avoided. But if the issue is important and the dissenter adamant, the gulf begins to widen. At first, the dissenter tries to exert all possible influence over the others, tries to bring the others around. In Albert Hirschman's compact terminology, this is the option of *voice*. Short of calling a press conference, this option can be exercised in several ways from simply grumbling to threatening to resign. But usually the individual gives voice to his dissatisfaction in a series of private confrontations like those of Muskie and Humphrey with Johnson. When these fail, as they usually do, he must face the possibility of resigning (or, as Hirschman calls it, exercising the option to *exit*). Resigning becomes a reasonable alternative as soon as voice begins to fail. The individual realizes that hours of sincere, patient argument have come to nothing. He realizes that his influence within the organization is waning, and so probably is his loyalty.

If he stays on, he risks becoming an organizational eunuch, an individual of no influence publicly supporting a policy against his will, judgment, personal value system, at times even his professional code.

As bleak as this prospect is, exit on matters of principle is still a distinctly uncommon response to basic institutional conflict. This is particularly true of American politics. In many nations with parliamentary systems, principled resignation from high office is common. But in the United States the concept of exit as a political act has never taken hold. The Walter Hickels are the exception. The last time a cabinet official left in protest and said why was when Labor Secretary Martin Durkin resigned because President Eisenhower refused to support his proposed amendments to the Taft-Hartley Act. As James Reston wrote recently in a postmortem on the Johnson Administration:

"One thing that is fairly clear from the record is that the art of resigning on principle from positions close to the top of American government has almost disappeared. Nobody quits now, as Anthony Eden and Duff Cooper left Neville Chamberlain's cabinet, with a clear and detailed explanation of why they couldn't be identified with the policy any longer. . . . Most [of those who stayed on] at the critical period of escalation gave to the President the loyalty they owed to the country. Some . . . are now wondering in private life whether this was in the national interest."

What accounts for our national reluctance to resign and our willingness, when forced to take the step, to settle for a "soft exit," without clamor, without a public statement of principle, and ideally without publicity? Tremendous institutional pressures and personal rationalizations work together to dissuade the dissident from exit in favor of voice. Most of us would much rather convince the boss or top group to see "reason" rather than quit. Resignation is defiant, an uncomfortable posture for most organization men (including politicians and academics). Worse, it smacks of failure, the worst of social diseases among the achievement-oriented. So instead of resigning, we reason to ourselves that the organization could go from bad to worse if we resigned. This may be the most seductive rationalization of all. Meanwhile, we have become more deeply implicated in the policy that we silently oppose, making extrication progressively more difficult. If resignation cannot be avoided, there are selfish reasons for doing it quietly. Most resignees would like to work again. Only Nader's Raiders love a blabbermouth. Speaking out is not likely to enhance one's marketability. A negative aura haunts the visibly angry resignee, while the individual who leaves a position ostensibly to return to business, family, teaching or research reenters the job market without any such cloud. Many resignees prefer a low profile simply because they are aware that issues change. Why undermine one's future effectiveness by making a noisy but ineffectual stand? However selfish the reasons, the organization reaps the major benefits when an individual chooses to resign quietly. A decorous exit conceals the underlying dissension that prompted the resignation in the first place. And the issue at contest is almost sure to be obscured by the speechmaking.

Like the Zen tea ceremony, resigning is a ritual, and woe to the man who fails to do it according to the rules. For example, when Fred Friendly resigned as President of CBS News in 1966 over the airing of Vietnam hearings, he sinned by releasing a news story *before* the Chairman of the Board, William S. Paley, could distribute his own release. Friendly writes in his memoir of this episode:

> Around two o'clock a colleague suggested that I should have called Paley, who was in Nassau, and personally read my letter [of resignation] to him over the phone. When I called Stanton to ask him if he had read my letter to the chairman, he said that he had just done so, and that Paley wanted me to call him. When I did, Paley wanted to know only if I had released my letter; when I told him that I had, all useful communication ceased. "You volunteered to me last week that you would not make a public announcement," he said. . . . The last thing the chairman said to me was: "Well, if you hadn't put out that letter, maybe we could still have done something." I answered that my letter was "after the fact, long after."

Paley's response is explicable only if we remember that the *fact* of resignation and the *reasons* behind it are subordinated in the organizational scheme to the issue of institutional face-saving. A frank resignation is regarded by the organization as an act of betrayal. (To some degree, this is, of course, an issue of personal face-saving. Those in power may wish for institutional harmony in part as a protection against personal criticism.)

Because a discreet resignation amounts to no protest at all, a soft exit lifts the opprobrium of organizational deviation from the resignee. When Dean Acheson bowed out as Under Secretary of the Treasury in 1933 after a dispute with F.D.R. over fiscal policy, his discretion was boundless and F.D.R. was duly appreciative. Some years later, when another official left with less politesse, sending the White House a sharp criticism of the President's policies, Roosevelt returned the letter with the tart suggestion that the man ought to "ask Dean Acheson how a gentleman resigns."

But "hard" or "soft," exit remains the option of last resort in organizational life. Remarkably, the individual who is deeply opposed to some policy often opts for public acquiescence and private frustration. He may continue to voice his opposition to his colleagues but they are able to neutralize his protest in various ways. Thus we see George Ball becoming the official devil's advocate of the Johnson Administration. As George E. Reedy writes:

> During President Johnson's Administration I watched George Ball play the role of devil's advocate with respect to foreign policy. The cabinet would meet and there would be an overwhelming report from Robert McNamara, another overwhelming report from Dean Rusk, another overwhelming report from McGeorge Bundy. Then five minutes would be set aside for George Ball to deliver his dissent, and because they expected him to dissent, they automatically discounted whatever he said. This strengthened them in their own convictions because the cabinet members could quite honestly say: "We heard both sides of this issue discussed." Well, they heard it with wax in their ears. I think that the moment you appoint an official devil's advocate you solidify the position he is arguing against.

One can hardly imagine a predicament more excruciating than Ball's. Often an individual in such conflict with the rest of his organization simply removes himself, if not physically then by shifting his concern from the issues to practical problems of management and implementation. He distracts himself. Townsend Hoopes suggests that this was the case with Robert McNamara. According to Hoopes, who was Under Secretary of the Air Force, there was growing evidence in the Autumn of 1967 that the President and McNamara were growing further and further apart in their attitudes toward escalating the Vietnam war. Hoopes saw in McNamara the fatigue and loneliness of a man "in deep doubt" as to the course the war was taking. But, writes Hoopes:

> *Owing to his own strict conception of loyalty to the President, McNamara found it officially necessary to deny all doubt and, by his silence, to discourage doubt in his professional associates and subordinates . . .* The result of McNamara's ambivalence, however, was to create a situation of dreamlike unreality for those around him. *His staff meetings during this period were entirely barren affairs: a technical briefing, for example, on the growing strength of air defenses around Hanoi, but no debate on what this implied for the U.S. bombing effort, and never the slightest disclosure of what the President or the Secretary of State might consider the broad domestic and international implications to be.* It was an atmosphere that worked to neutralize those who were the natural supporters of his concerns about the war. (Italics are for emphasis.)

What Hoopes describes is ethical short-circuiting. Conflict-torn McNamara busies himself with the minutiae of war planning because lists of numbers and cost estimates have a distracting if illusory moral neutrality. According to Hoopes, toward the end of McNamara's tenure, the despairing Secretary stopped questioning the military and political significance of sending 206,000 more troops into Indochina and concentrated in the short time he had on the logistical problems of getting them to the port of debarkation safely and efficiently.

One sees a remarkably similar displacement of energy from moral or political concerns to managerial or technological ones in the career of Albert Speer. I do not mean to label McNamara a Fascist by literary association. But the pages of *Inside the Third Reich* reveal that Speer dealt with ambivalence brought on by intense organizational stress in a remarkably similar way. Speer did not allow his growing personal reservations about Hitler to interfere with his meticulous carrying out of administrative duties. Speer kept the Nazi war machine running in high gear and increasingly productive until 1945. As Eugene Davidson writes: "A man like Speer, working with blueprints, ordering vast projects, is likely to exhaust himself in manipulation, in transforming the outer world, in carrying out production goals with all the means at hand."

Whether such activity exhausts an individual to the point of moral numbness is questionable, but certainly the nature of the large organization makes it possible for a McNamara or an Albert Speer or an Ellsberg (while at Rand), for that matter, to work toward an ultimately immoral end without an immediate sense of personal responsibility or guilt. Organizations are by definition systems of increased differentiation and specialization, and, thus, the morality of the organization is the morality of segmented acts. As Charles Reich wrote in *The New Yorker,* "A scientist who is doing his specialized duty to further research and knowledge develops a substance called napalm. Another specialist makes policy in the field of our nation's foreign affairs. A third is concerned with the most modern weaponry. A fourth manufactures what the defense authorities require. A fifth drops napalm from an airplane where he is told to do so." In this segmented environment, any one individual can easily develop tunnel vision, concentrating on the task at hand, completing his task with a sense of accomplishment, however sinister the collective result of all these individual jobs well done. This segmented structure characteristic of all large organizations encourages indifference and evasion of responsibility. A benefit of membership in such an organization is insurance against the smell of burning flesh. Speer, for example, still does not seem particularly troubled by the horrors of slave labor in his wartime munitions plants even when making his unique public confession.

Speer reports that it never occurred to him to resign even though he was aware of what his loss would do to hasten the end of Hitler's regime. Faced with a much more subtle and complex situation, McNamara seriously considered resigning, according to Hoopes. But that he did not do so in 1967 when his doubts were so oppressive is remarkable. Hoopes provides a fascinating clue to McNamara's reluctance to resign or even to voice his uneasiness in any except the most private audiences with the President. In the following short portrait by Hoopes in his book *The Limits of Intervention,* we see McNamara wrestling with an ingrained organizational ethic stronger than his own intelligence and instinct:

> Accurately regarded by the press as the one moderate member of the inner circle, he continued to give full public support to the Administration's policy, including specific endorsement of successive manpower infusions and progressively wider and heavier bombing efforts. Inside the Pentagon he seemed to discourage dissent among his staff associates by the simple tactic of being unreceptive to it; he observed, moreover, so strict a sense of privacy in his relationship with the President that he found it virtually impossible to report even to key subordinates what he was telling the President or what the President was saying and thinking. . . . All of this seemed to reflect a well-developed philosophy of executive management relationships, derived from his years in industry; its essence was the belief that a busy, overworked chairman of the board should be spared the burden of public differences among his senior vice-presidents. Within such a framework, he could argue the case for moderation with the President—privately, selectively, and intermittently. But the unspoken corollary seemed to be that, whether or not his counsel of moderation were followed, there could arise no issue or difference with President Johnson sufficient to require his resignation—whether to enlighten public opinion or avoid personal stultification. It was this corollary that seemed of doubtful applicability to the problems and obligations of public office. *McNamara gave evidence that he had ruled out resignation because he believed that the situation would grow worse if he left the field to Rusk, Rostow, and the Joint Chiefs. But also because the idea ran so strongly against the grain of his temperament and his considered philosophy of organizational effectiveness.*

Does this mean that McNamara would not resign because quitting violated some personal notion of honor? Or does it mean that he believed that dissent and "organizational effectiveness" are negatively correlated? I suspect that the latter is closer to the truth. Like any other corporation president, McNamara was raised on organizational folklore. One of the central myths is that the show of unanimity is always desirable. That this belief is false and even dangerous does not limit its currency. Yes, there are times when discretion is required. Clearly organizations should not fight constantly in public. But what is the gain of forbidding at all costs and at all times any emotional give-and-take between colleagues? A man has an honest difference of opinion with the organizational powers. Why must he be silenced or domesticated or driven out so that the public can continue to believe—falsely— that organizational life is without strife? And yet organizations continue to assume the most contrived postures in order to maintain the illusion of harmony. Postures like lying to the public.

Our inability to transcend the dangerous notion that we don't wash our dirty linen in public verges on the schizophrenic. It implies that dissent is not only bad but that our public institutions such as government are made up not of men but saints who never engage in such vulgar and offensive activities. Thus, government strives to be regarded as a hallowed shrine where, as George Reedy reports from his experience as White House press secretary under President Johnson, "the meanest lust for power can be sanctified and the dullest wit greeted with reverential awe." In fact, organizations, including governments, are vulgar, sweaty, plebeian; if they are to be viable, they must create an institutional environment where a fool can be called a fool and all actions and motivations are duly and closely scrutinized for the inevitable human flaws and failures. In a democracy, meanness, dullness, and corruption are always amply represented. They are not entitled to protection from the same rude challenges that such qualities must face in the "real" world. When banal politeness is assigned a higher value than accountability or truthfulness, the result is an Orwellian world where the symbols of speech are manipulated to create false realities.

"Loyalty" is often given as a reason or pretext for muffling dissent. A variation on this is the claim that candor "gives comfort to the enemy." Ellsberg's national loyalty was repeatedly questioned in connection with his release of the so-called Pentagon Papers. In the first three installments of the document as run in The *Times,* practically nothing that wasn't well known was revealed. A few details, an interesting admission or two, but basically nothing that had not come to light earlier in other less controversial articles and books on the Indochina war. But government officials trying to suppress the publication of the classified material chose to make much of the "foreign consequences" of its release. "You may rest assured," a government official was quoted as saying by the Buffalo *Evening News,* "that no one is reading this series any more closely than the Soviet Embassy."

All of the foregoing pressures against registering dissent can be subsumed under the clumsy label of "loyalty." In fact, they represent much more subtle personal and organizational factors including: deep-rooted psychological dependence, authority problems, simple ambition, co-optive mechanisms (the "devil's advocacy" technique), pressure to be a member of the club and fear of being outside looking in, adherence to the myth that gentlemen settle their differences amicably and privately, fear of disloyalty in the form of giving comfort to "the enemy," and, very often, that powerful Prospero-aspiration: the conviction that one's own "reasonable" efforts will keep things from going from bad to worse.

There is a further broad cultural factor that must be considered before the other defenses against exit can be understood. It simply doesn't make sense for a man as intelligent and analytically sophisticated as our nation's Number One Problem Solver, Robert McNamara, to delude himself that he couldn't quit because "duty called." Duty to whom? Not to his own principles? Nor, as he saw it, to the nation's welfare. McNamara's real loyalty was to the code of the "organizational society" in which most of us live out our entire active careers. Ninety percent of the employed population of this country works in formal organizations. Status, position, a sense of competence and accomplishment are all achieved in our culture through belonging to these institutions. What you *do* determines, to a large extent, what you *are*. "My son, the doctor" is not only the punch line of a thousand Jewish jokes. It is a neat formulation of a significant fact about our culture. Identification with a profession or other organization is a real-life passport to identity, to selfhood, to self-esteem. You are what you do, and work in our society (as in all other industrialized societies) is done in large, complex, bureaucratic structures. If one leaves the organization, particularly with protest, one is nowhere, like a character in a Beckett play, without role, without the props of office, without ambience or setting.

In fact, a few more resignations would be good for individual consciences and good for the country. Looking back, veteran diplomat Robert Murphy could recall only one occasion when he thought he should have resigned. The single instance was the Berlin Blockade of 1948–49, which he thought the U.S. should have challenged more vigorously. "My resignation almost certainly would not have affected events," he wrote in regret, "but if I had resigned, I would feel better today about my own part in that episode." *Time* magazine, from which Murphy's quotation was taken, goes on to say in its essay:

> In the long run, the country would probably feel better, too, if a few more people were ready to quit for their convictions. It might be a little unsettling. But it could have a tonic effect on American politics, for it would give people the assurance that men who stay truly believe in what they are doing.

My own resignation was a turning point. The decision represented the first time in many years of organizational life that I had been able to say, No, I cannot allow myself to be identified with that particular policy, the first time I had risked being an outsider rather than trying to work patiently within the system for change. Many factors entered into the decision but in the last analysis my reason for resigning was an intensely personal one. I did not want to say, a month or two months after the police came onto campus, "Well, I was against that move at the time." I think it is important for everyone in decision-making positions in our institutions to speak out. And if we find it impossible to continue on as administrators because we are at total and continuous odds with institutional policy, then I think we must quit and go out shouting. The alternative is petit-Eichmannism, and that is too high a price.

Note

[1]For many reasons, notably my decision to retain another administrative position while resigning the acting post. The distinction between the positions was clear only to other members of the administration, and the public generally interpreted my equivocal exit as a halfhearted protest.

11
Followership

*Like "When to Resign," this short piece is about
doing the right thing. Subordinates sometimes pay
the ultimate institutional price for candor, but that
doesn't relieve them of the obligation to tell their
leaders what they may not want to hear. This piece
is also a reminder that morality is not solely an executive function.*

It is probably inevitable that a society as star-struck as ours
should focus on leaders in analyzing why organizations succeed
or fail. As a long-time student and teacher of management, I, too,
have tended to look to the men and women at the top for clues on
how organizations achieve and maintain institutional health. But
the longer I study effective leaders, the more I am convinced of
the under-appreciated importance of effective followers.

What makes a good follower? The single most important characteristic may well be a willingness to tell the truth. In a world of
growing complexity, leaders are increasingly dependent on their
subordinates for good information, whether the leaders want to
hear it or not. Followers who tell the truth, and leaders who listen
to it, are an unbeatable combination.

Movie mogul Samuel Goldwyn seems to have had a gut-level
awareness of the importance of what I call "effective backtalk"
from subordinates. After a string of box-office flops, Mr. Goldwyn
called his staff together and told them: "I want you to tell me exactly what's wrong with me and M.G.M., even if it means losing
your job."

Although Mr. Goldwyn wasn't personally ready to give up the
ego-massaging presence of "yes men," in his own gloriously gar-
bled way he acknowledged the company's greater need for a staff
that speaks the truth.

Like portfolios, organizations benefit from diversity. Effective leaders resist the urge to people their staffs only with others who look or sound or think just like themselves, what I call the doppelgänger, or ghostly-double, effect. They look for good people from many molds, and then they encourage them to speak out, even to disagree. Aware of the pitfalls of institutional unanimity, some leaders wisely build dissent into the decision-making process.

Organizations that encourage thoughtful dissent gain much more than a heightened air of collegiality. They make better decisions. In a recent study, Rebecca A. Henry, a psychology professor at Purdue University, found that groups were generally more effective than individuals in making forecasts of sales and other financial data. And the greater the initial disagreement among group members, the more accurate the results. "With more disagreement, people are forced to look at a wider range of possibilities," Ms. Henry said.

Like good leaders, good followers understand the importance of speaking out. More important, they do it. Almost 30 years ago, when Nikita Khruschev came to America, he met with reporters at the Washington Press Club. The first written question he received was: "Today you talked about the hideous rule of your predecessor, Stalin. You were one of his closest aides and colleagues during those years. What were you doing all that time?" Khruschev's face grew red. "Who asked that?" he roared. No one answered. "Who asked that?" he insisted. Again, silence. "That's what I was doing," Mr. Khruschev said.

Even in democracies, where the only gulag is the threat of a pink slip, it is hard to disagree with the person in charge. Several years ago TV's John Chancellor asked former Presidential aides how they behaved on those occasions when the most powerful person in the world came up with a damned fool idea. Several of the aides admitted doing nothing. Ted Sorenson revealed that John F. Kennedy could usually be brought to his senses by being told, "That sounds like the kind of idea Nixon would have."

Quietism, as a more pious age called the sin of silence, often costs organizations—and their leaders—dearly. Former President Ronald Reagan suffered far more at the hands of so-called friends who refused to tell him unattractive truths than from his ostensible enemies.

Nancy Reagan, in her recent memoir, "My Turn," recalls chiding then–Vice President George Bush when he approached her, not the President, with grave reservations about White House chief of staff Donald Regan.

"I wish you'd tell my husband," the First Lady said. "I can't be the only one who's saying this to him." According to Mrs. Reagan, Mr. Bush responded, "Nancy, that's not my role."

"That's exactly your role," she snapped.

Nancy Reagan was right. It is the good follower's obligation to share his or her best counsel with the person in charge. And silence—not dissent—is the one answer that leaders should refuse to accept. History contains dozens of cautionary tales on the subject, none more vivid than the account of the murder of Thomas à Becket. "Will no one rid me of this meddlesome priest?" Henry II is said to have muttered, after a contest of wills with his former friend.

The four barons who then murdered Becket in his cathedral were the antithesis of the good followers they thought themselves to be. At the risk of being irreverent, the right answer to Henry's question—the one that would have served his administration best—was "No," or at least, "Let's talk about it."

Like modern-day subordinates who testify under oath that they were only doing what they thought their leader wanted them to do, the barons were guilty of remarkable chutzpah. Henry failed by not making his position clear and by creating an atmosphere in which his followers would rather kill than disagree with him. The barons failed by not making the proper case against the king's decision.

Effective leaders reward dissent, as well as encourage it. They understand that whatever momentary discomfort they experience as a result of being told from time to time that they are wrong is more than offset by the fact that reflective backtalk increases a leader's ability to make good decisions.

Executive compensation should go far toward salving the pricked ego of the leader whose followers speak their minds. But what's in it for the follower? The good follower may indeed have to put his or her job on the line in the course of speaking up. But consider the price he or she pays for silence. What job is worth the enormous psychic cost of following a leader who values loyalty in the narrowest sense?

Perhaps the ultimate irony is that the follower who is willing to speak out shows precisely the kind of initiative that leadership is made of.

12

Ethics Aren't Optional

At their heart, questions of ethics are questions of character, the overridingly important criterion by which leaders must be judged. All good leaders (in both senses of the word) manage to keep three balls in the air: competence, drive, and integrity. Even when one side wasn't accusing the other of lacking character, it was a major issue in the 1992 presidential campaign—and rightly so. As Sophocles said, "Power reveals the man." I must say I am more hopeful today about our culture's ethics than when I wrote this piece a few years ago.

There seem to be no innocents left in America. Ollie North claims to be a patriot, but his patriotism has resulted in a home security system and a numbered Swiss bank account. Teen-agers peddle drugs on the streets in East Los Angeles. Self-proclaimed men of God recite the Ten Commandments on Sundays and break them the other six days of the week. Spying for dollars is our latest growth industry, and Wall Street looks more and more like a branch of Sing Sing.

In the late 1960s, yippies were political activists. In the 1980s, yippies are young indicted professionals, while yuppies, our grand acquisitors, consumed by consumption, turn out to be unindicted professionals—those who haven't got caught yet.

Young children are pushed into boutique nursery schools, where excellence is measured by the cut of one's Polo shirt. Teen-agers drive VW Cabriolets, and are pressured to score stupendously on their SATs so they can go to Brown with Amy Carter and Cosima Von Bulow, and graduate to investment banking where the players use real money and jail is the only limit.

Kids no longer dream of going to the moon, or making a better mousetrap. They dream of money and they know that the best things in life are VCRs, cellular telephones, Beemers, dinner at the Quilted Giraffe or Rebecca's. They don't vote, of course, believing that politics are obsolete, along with politicians.

Politician Gary Hart made himself obsolete by denying that he was a womanizer, then womanizing and claiming he hadn't. When faced with further evidence of his womanizing, he quit the race for the Democratic Presidential nomination and suggested that he was too good for us. The American people seemed as confused as Hart himself as to what he did wrong. Was he brought down because he allegedly committed adultery or because he allegedly lied? Or was it because he was too proud and too arrogant? Or was he simply guilty of bad judgment?

The traditionalists objected most to his alleged adultery. The more modern among us could excuse Hart's purported sin of adultery on the grounds that what a man does in the bedroom is no one's business. But we were bothered by his apparent lies.

The hippest and most contemporary citizens dismissed the charges of both adultery and lying, because after all everyone who is anyone does both, but they saw Hart's bad judgment as his fatal flaw. In other words, they weren't bothered by what he did, but that he got caught. So far, I haven't run into anyone who was disturbed by Hart's hubris. Once upon a time, pride was a sin. Now it's a virtue.

Our national confusion over Hart's alleged mistakes vividly demonstrates a startling ethical decline. It is not simply that more of us are engaging in unethical behavior, it is because more and more we are unwilling or unable to identify or define what constitutes unethical behavior.

After the penultimate Wall Street trader Ivan Boesky was nailed, a TV news crew went into a Wall Street bar and interviewed some young traders. Each and every one expressed admiration for Boesky and contempt for the Securities and Exchange Commission. Earlier, when four of their own were caught playing games that were too fast and loose even for Wall Street, the disgraced young traders were more censured than pitied. Winning isn't everything, it's the only thing, and getting caught is for losers. And, as one market analyst said, it isn't a bull market or a bear market, "it's a pig market."

After another round of arrests on the Street, an investment banker told the *New York Times* that the sight of their colleagues in handcuffs "put the fear of God in everybody." Such late-inning invocations are, of course, S.O.P. for white-collar felons, as we saw in the wake of Watergate. And why not? Almost anyone would rather wear a halo than handcuffs. But, as the recent revelations about Jim and Tammy Bakker and their PTL (Praise the Lord) Club show, the church is no holier than Wall Street, and at least as profitable.

Ollie North's exercise in patriotism for profit, the games other White House notables like Michael Deaver and Attorney General Meese are charged with playing, Wall Street's dirty dozen, Hart's fall, the Bakkers' highly secular adventures are the latest manifestations of a social crisis of enormous proportions.

In this highly materialistic nation, the prevailing ethic is, at best, pragmatic, and, at worst, downright dishonest. It's every man for himself, and never mind God, country or anything else. There seems to be no such thing as the common good or the public interest. Only self-interest. That old entrepreneurial spirit that Ronald Reagan admires so ardently is running amuck, and the country is coming unstuck.

Ted Turner buys MGM and guts it. GE gobbles up RCA while the airlines feed on each other. TV evangelists squeeze big bucks out of believers, and Wall Street traders and Washington patriots peddle their services to the highest bidders. The rich get richer and the poor get poorer. And the federal deficit gets bigger. As the poet William Butler Yeats said in another time and place, "the center is not holding."

It is time, then, to face this ethical deficit or America will end in shambles. Ethics and conscience aren't optional. They are the glue that binds society together—the quality in us that separates us from cannibals. Without conscience and ethics, talent and power amount to nothing.

13
Change: The New Metaphysics

How change happens and how to make it happen.

Change is the metaphysics of our age. Everything is in motion. Everything mechanical has evolved, become better, more efficient, more sophisticated. In this century, automobiles have advanced from the Model T to the BMW, Mercedes, and Rolls Royce. Meanwhile, everything organic—from ourselves to tomatoes—has devolved. We have gone from such giants as Teddy Roosevelt, D. W. Griffith, Eugene Debs, Frank Lloyd Wright, Thomas Edison and Albert Michelson to Yuppies. Like the new tomatoes, we lack flavor and juice and taste. Manufactured goods are far more impressive than the people who make them. We are less good, less efficient, and less sophisticated with each passing decade.

People in charge have imposed change rather than inspiring it. We have had far more bosses than leaders, and so, finally, everyone has decided to be his or her own boss. This has led to the primitive, litigious, adversarial society we now live in. As the newscaster in the movie *Network* said, "I'm mad as hell, and I'm not going to take it anymore."

What's going on is a middle-class revolution. The poor in America have neither the time nor the energy to revolt. They're just trying to survive in an increasingly hostile world. By the same token, the rich literally reside above the fray—in New York penthouses, Concordes, and sublime ignorance of the world below. The middle class aspires to that same sublime ignorance.

A successful dentist once told me that people become dentists to make a lot of money fast and then go into the restaurant business or real estate, where they will really make money. Young writers and painters are not content to practice their craft and perfect it. Now they want to see and be seen, wheel and deal, and they are as obsessed with the bottom line as are IBM executives. The deal for the publication of a book is far more significant than the book itself, and the cover of *People* magazine is more coveted than a good review in the *New York Times*. The only unions making any noise now are middle-class unions. Professors who once professed an interest in teaching are now far more interested in deals—for the book, the TV appearance, the consulting job, the conference in Paris—leaving teaching to assistants.

When everyone is his or her own boss, no one is in charge, and chaos takes over. Leaders are needed to restore order, by which I mean not obedience but progress. It is time for us to control events rather than be controlled by them.

Avenues of Change

Change occurs in several ways.

- *Dissent and conflict.* We have tried dissent and conflict and have merely become combative. In corporations, change can be mandated by the powers that be. But this leads inevitably to the escalation of rancor. We are perpetually angry now, all walking around with chips on our shoulders.

- *Trust and truth.* Positive change requires trust, clarity and participation. Only people with virtue and vision can lead us out of this bog and back to the high ground, doing three things: (1) gaining our trust; (2) expressing their vision clearly so that we all not only understand but concur; and (3) persuading us to participate.

- *Cliques and cabals.* The cliques have the power, the money and the resources. The cabals, usually younger and always ambitious, have drive and energy. Unless the cliques can co-opt the cabals, revolution is inevitable. This avenue, too, is messy. It can lead to either a stalemate or an ultimate victory for the cabals, if for no other reason than they have staying power.

- *External events.* Forces of society can impose themselves on the organization. For example, the auto industry was forced to change its ways and its products, both by government regulation and by foreign competition. In the same way, student activists forced many universities to rewrite their curricula and add black studies and women's studies programs. Academicians are still debating both the sense and the efficacy of such programs, as they have altered not only what students learn but how they learn it.

• *Culture or paradigm shift.* The most important avenue of change is culture or paradigm. In *The Structure of Scientific Revolution,* Thomas Kuhn notes that the paradigm in science is akin to a zeitgeist or climate of opinion that governs choices. He defines it as "the constellation of values and beliefs shared by the members of a scientific community that determines the choice, problems which are regarded as significant, and the approaches to be adopted in attempting to solve them." The people who have revolutionized science have always been those who have changed the paradigm.

Innovators and Leaders

People who change not merely the content of a particular discipline but its practice and focus are not only innovators but leaders. Ralph Nader, who refocused the legal profession to address consumer problems, was such a person. Betty Friedan, in truthfully defining how women lived, inspired them to live in different ways.

It is not the articulation of a profession or organization's goals that creates new practices but rather the imagery that creates the understanding, the compelling moral necessity for the new way. The clarity of the metaphor and the energy and courage its maker brings to it are vital to its acceptance. For example, when Branch Rickey, general manager of the Brooklyn Dodgers, decided to bring black players into professional baseball, he chose Jackie Robinson, a paragon among players and among men.

How do we identify and develop such innovators? How do we spot new information in institutions, organizations and professions? Innovators, like all creative people, see things differently, think in fresh and original ways. They have useful contacts in other areas; they are seldom seen as good organization men or women and often viewed as mischievous troublemakers. The true leader not only is an innovator but makes every effort to locate and use other innovators in the organization. He or she creates a climate in which conventional wisdom can be challenged and one in which errors are embraced rather than shunned in favor of safe, low-risk goals.

In organizations, people have norms, values, shared beliefs and paradigms of what is right and what is wrong, what is legitimate and what is not, and how things are done. One gains status and power through agreement, concurrence, and conformity with these paradigms. Therefore, both dissent and innovation are discouraged. Every social system contains these forces for conservatism, for maintaining the status quo at any cost, but it must also contain means for movement, or it will eventually become paralyzed.

Basic changes take place slowly because those with power typically have no knowledge, and those with knowledge have no power. Anyone with real knowledge of history and the world as it is today could redesign society, develop a new paradigm in an afternoon, but turning theory into fact could take a lifetime.

Still, we have to try because too many of our organizations and citizens are locked into roles and practices that simply do not work. True leaders work to gain the trust of their constituents, communicate their vision lucidly, and thus involve everyone in the process of change. They then try to use the inevitable dissent and conflict creatively and positively, and out of all that, sometimes, a new paradigm emerges.

A Harris poll showed that over 90 percent of the people polled would change their lives dramatically if they could, and they ranked such intangibles as self-respect, affection and acceptance higher than status, money and power. They don't like the way they live now, but they don't know how to change. The poll is evidence of our need for real leaders and should serve as impetus and inspiration to potential leaders and innovators. If such people have the will to live up to their potential—and the rest of us have the gumption to follow them—we might finally find our way out of this bog we're in.

Avoiding Disaster during Change

Constant as change has been and vital as it is now, it is still hard to effect, because the sociology of institutions is fundamentally antichange. Here, then, are 10 ways to avoid disaster during periods of change—any time, all the time—except in those organizations that are dying or dead.

1. *Recruit with scrupulous honesty.* Enthusiasm or plain need often inspires recruiters to transmogrify visible and real drawbacks and make them reappear as exhilarating challenges. Recruiting is, after all, a kind of courtship ritual. The suitor displays his or her assets and masks his or her defects. The recruit, flattered by the attention and the promises, does not examine the proposal thoughtfully. He or she looks forward to opportunities to be truly creative and imaginative and to support from the top.

Inadvertently, the recruiter has cooked up the classic recipe for revolution as suggested by Aaron Wildavsky: "Promise a lot; deliver a little. Teach people to believe they will be much better off, but let there be no dramatic improvement. Try a variety of small programs but marginal in impact and severely underfinanced. Avoid any attempted solution remotely comparable in size to the dimensions of the problem you're trying to solve."

2. *Guard against the crazies.* Innovation is seductive. It attracts interesting people. It also attracts people who will distort your ideas into something monstrous. You will then be identified with the monster and be forced to spend precious energy combating it. Change-oriented managers should be sure that the people they recruit are change agents but not agitators. It is difficult sometimes to tell the difference between the innovators and the crazies. Eccentricities and idiosyncrasies in change agents are often useful and valuable. Neurosis isn't.

3. *Build support among like-minded people,* whether or not you recruited them. Change-oriented administrators are particularly prone to act as though the organization came into being the day they arrived. This is a delusion, a fantasy of omnipotence. There are no clean slates in established organizations. A new CEO can't play Noah and build the world anew with a handpicked crew of his or her own. Rhetoric about new starts is frightening to those who sense that this new beginning is the end of their careers. There can be no change without history and continuity. A clean sweep, then, is often a waste of resources.

4. *Plan for change from a solid conceptual base.* Have a clear understanding of how to change as well as what to change. Planning changes is always easier than implementing them. If change is to be permanent, it must be gradual. Incremental reform can be successful by drawing on a rotating nucleus of people who continually read the data provided by the organization and the society in which it operates for clues that it's time to adapt. Without such critical nuclei, organizations cannot be assured of continued self-renewal. Such people must not be faddists but must be hypersensitive to ideas whose hour has come. They also know when ideas are antithetical to the organization's purposes and values and when they will strengthen the organization.

5. *Don't settle for rhetorical change.* Significant change cannot be decreed. Any organization has two structures: one on paper and another that consists of a complex set of intramural relationships. A good administrator understands the relationships and creates a good fit between them and any planned alterations. One who gets caught up in his or her own rhetoric almost inevitably neglects the demanding task of maintaining established constituencies and building new ones.

6. *Don't allow those who are opposed to change to appropriate basic issues.* Successful change agents make sure that respectable people are not afraid of what is to come and that the old guard isn't frightened at the prospect of change. The moment such people get scared is the moment they begin to fight dirty. They not only have some built-in clout, they have tradition on their side.

7. *Know the territory.* Learn everything there is to know about the organization and about its locale, which often means mastering the politics of local chauvinism, along with an intelligent public relations program. In Southern California, big developers are constantly being blindsided by neighborhood groups because they have not bothered to acquaint the groups with their plans. Neighborhood groups often triumph, forcing big changes or cancellations. They know their rights and they know the law, and the developers haven't made the effort to know them.

8. *Appreciate environmental factors.* No matter how laudable or profitable or imaginative, a change that increases discomfort in the organization is probably doomed. Adding a sophisticated new computer system is probably a good thing, but it can instantly be seen as a bad thing if it results in overcrowded offices.

9. *Avoid future shock.* When an executive becomes too involved in planning, he or she frequently forgets the past and neglects the present. As a result, before the plan goes into effect, employees are probably already opposed to it. They, after all, have to function in the here and now, and if their boss's eye is always on tomorrow, he or she is not giving them the attention and support they need.

10. *Remember that change is most successful when those who are affected are involved in the planning.* This is a platitude of planning theory, but it is as true as it is trite. Nothing makes people resist new ideas or approaches more adamantly than their belief that change is being imposed on them.

The problems connected with innovation and change are common to every modern bureaucracy. University, government and corporation all respond similarly to challenge and to crisis, with much the same explicit or implicit codes, punctilios and mystiques.

Means must be found to stimulate the pursuit of truth—that is, the true nature of the organization's problems—in an open and democratic way. This calls for classic means: an examined life, a spirit of inquiry and genuine experimentation, a life based on discovering new realities, taking risks, suffering occasional defeats, and not fearing the surprises of the future. The model for truly innovative organizations in an era of constant change is the scientific model. As scientists seek and discover truths, so organizations must seek and discover their own truths—carefully, thoroughly, honestly, imaginatively, and courageously.

14

Meet Me in Macy's Window

It is hard, even today, to recall the wholesale mistrust of public officials that characterized the Watergate era. (There were bumper stickers that urged "Impeach Somebody!") One of the results of that dark time was an urgent call for greater organizational candor. This piece, first published in 1975, argues that complete candor is a false solution to a real problem—that of trust. The essay calls for balancing the legitimate needs for openness and confidentiality in the interest of effectiveness.

The British Foreign Office gives its fledgling diplomats three cardinal rules of behavior: (1) never tell a lie, (2) never tell the whole truth, and (3) never miss a chance to go to the bathroom. An old Tammany boodler, who disliked leaving any traces of his dealings, had a terser rule: "Don't write. Send word."

Both sets of rules, I fear, are likely to become more and more a tacit standard of conduct for those who, in the post-Watergate climate of suspicion, share the hazardous privilege of running large organizations—including, in my own case, the nation's second largest urban multiversity.

Never have the American people felt such universal distrust of their presumed leaders—whether in government, the law, the clergy, or education. Year after year of calculated deception over Vietnam, compounded by the conspiracy, skullduggery, and chicanery of Watergate, have left them trusting almost no one in authority.

Consider a recent Gallup survey in which college students were asked to rate the honesty and ethical standards of various groups: political officeholders (only 9 percent rated "very high") were eclipsed only by advertising men (6 percent), lawyers rated

40 percent, journalists 49 percent. I am proud that college teachers rated highest (70 percent), but since college presidents were not included, I can't seek shelter under that umbrella. Ralph Nader got a higher rating than President Gerald Ford, Henry Kissinger, or Ted Kennedy. Labor leaders came out worse than business executives—only 19 percent were rated high, versus 20 percent for the latter.

In short, virtually all leaders are in the doghouse of suspicion. And the understandable reaction to all these credibility gaps is creating a growing insistence that every public act, of whatever public institution, be conducted, as it were, in Macy's window.

Some symptoms:

• "Sunshine laws" have now been passed by numerous states, prohibiting closed meetings at any federally supported institution. Hawaii has even made it a crime to hold a private meeting of *any* sort without giving advance notice.

• The Buckley Amendment requires that personnel records in institutions with federal support (particularly those concerning students) be open to inspection by the person concerned.

• The Freedom of Information Act, first passed in 1967, and recently strengthened over the President's veto by amendments that became effective last February 19, requires that most records of federal agencies be provided to anyone upon request.

The intended purpose of all such measures is wholesome. It is to create a standard, for all public business, of what Woodrow Wilson called "open covenants openly arrived at." I believe wholeheartedly in such a purpose. Over many years of consulting, teaching, and writing on the achievement of organizational goals (for all organizations, but particularly those of business and government), I have always stressed the importance of openness.

I have argued that goals will be achieved effectively almost in proportion to the extent that the organization can achieve a climate where members can level with one another in open and trusting interpersonal relationships. I believe this—because denial, avoidance, or suppression of truth will ultimately flaw the decision-making—and in the case of business, the bottom line as well.

So—I dislike secrecy. I think Luke was right when he wrote, "Nothing is secret, that shall not be made manifest." I believe Emerson's law of compensation: "In the end, every secret is told, every crime is punished, every virtue rewarded, in silence and certainty."

At the same time, I am convinced, as a practical administrator, that these well-intended goldfish-bowl rules will have unintended results worse than the evils they seek to forestall. They are likely to produce more secrecy, not less (only more carefully concealed), and on top of it, so hamstring already overburdened administrators as to throw their tasks into deeper confusion.

For secrecy is one thing. Confidentiality is another. No organization can function effectively without certain degrees of confidentiality in the proposals, steps, and discussions leading up to its decisions—which decisions should then, of course, be open, and generally will be.

An amusing case in point. The Nixon Administration moved heaven and earth seeking to restrain, perhaps even imprison, *New York Times* editors in their determination to publish the Pentagon papers. The *Times* won the right from the Supreme Court (under some continuing criminal risk) to resume publishing these assertedly "secret" studies of Vietnam War decisions. Yet the editors themselves surrounded their preparation of these stories with a secrecy and security the Pentagon might envy—renting a secret suite of hotel rooms, swearing each member of a small secret staff to total secrecy, confining them for weeks almost like prisoners, restricting their communications to an elite handful with "need to know," and setting the stories themselves on sequestered, closely guarded typesetting machines. Thus the ultimate challenge to "official" secrecy was performed in ultimate "private" secrecy.

What the *Times* editors knew, of course, was what every decision-maker knows instinctively. The mere fact of discussions becoming known, at the wrong stage of the procedure, can prevent a desirable decision from ultimately being carried out.

We have seen this happen in the case of the long, arduous, confidential negotiations Secretary of State Kissinger was making with the Soviets to tie trade concessions to larger, mutually agreed quotas of emigration for Soviet Jews.

He had already obtained, through quiet negotiation, large but unstipulated expansions of the actual numbers of émigrés, who began arriving in Israel by the thousands. He obtained similar agreement to larger expansions. But zealous Senatorial advocates of larger emigration demanded that all this be put in Macy's window—be publicly recorded, that the Soviets publicly confirm what they were privately conceding. The outcome was to rupture the progress that had already been gained in emigration.

On a less cosmic level, some experiences of my own bring home how vital confidentiality can be in determining whether or not ultimately "open decisions openly arrived at" can be made at all.

Case 1:

Shortly after I became president of the University of Cincinnati, of which the city's largest hospital (General) is a part, a U.S. Senator announced an investigation of whether whole-body radiation, carried out at General on terminal cancer patients, constituted "using human beings as guinea pigs." The charges were totally false, but there were some awkward aspects in the way the whole thing had been handled that caused me to investigate (privately) the reasons.

This was on the eve of a Hamilton County election absolutely crucial to the hospital, on which thousands of the poor rely for treatment. It was far from sure whether a major bond levy for General Hospital would pass or fail. It did pass, but during three critical weeks I had either to evade all questions, or fuzz my answers, relating to my own and to the Senator's investigation. I never lied. I never told the whole truth. I often went to the bathroom.

Case 2:

Our university, which began as a city-funded municipal college and to which the City of Cincinnati still contributes $4.5 million of its annual $140 million budget, now draws the bulk of its funding from the state. But it is not a full state institution like Ohio State. If we *were* fully state-affiliated, we would receive sufficient extra funds to meet a worsening financial crisis. The possibility of such affiliation therefore not only *needs* to be considered but *has* to be considered—I would be derelict in my duty to do otherwise.

But if we decided to seek full state status, timing was very important—since it would involve not only action by the legislature, but a change in the city's charter. Even more important, I learned to my sorrow, was confidentiality.

One of our state senators, preparing for a TV interview, asked me if it was all right for him to say that the university was "considering" such a move. I said, "Certainly," since obviously I had to consider it. By night this statement of the obvious was "big" news flashing across my TV screen. By morning local and state politicians were making a pro-and-con beanbag of the question, and by then the furor was such that it was difficult even to weigh or discuss the problem on its merits. Happily, that frenetic period has now passed and the question is being calmly and thoughtfully debated—but I learned a lesson.

Case 3:

Last year a group of black graduate students made charges of racism against their college faculty. I met with this group and heard out their grievances. I told them that, if the faculty would agree, I would ask a blue-ribbon panel of distinguished local citizens, including two black leaders, to investigate and report on the matter.

That was Wednesday. On the next day, Thursday, the dean of the College had arranged to meet with the faculty. The plan was to make this proposal for such a committee. I had no reason to think the faculty would object.

But by late that Wednesday afternoon the *Cincinnati Post* was blazoning the entire story—the protest meeting, my proposal to the students, tomorrow's meeting arranged with the faculty, etc. Apparently, the protesters had "leaked" the details of our meeting, on the assumption that it would further their cause. The opposite happened. The faculty were irritated by reading about arrangements they had not been consulted about. By the time I *could* consult them, they were sufficiently angry to vote down the whole proposal of an outside committee. Werner Heisenberg's "uncertainty principle" affects human as well as atomic relations:

the mere act of observing a process (publicly) *can impede the process itself.*

So—it is certainly clear in my own mind that there are times when confidentiality is a necessary prerequisite to *public* decisions for the *public* benefit. But when one asks, or is asked, where this desirable good blends into the undesirable evil of secrecy—for secrecy's own sake, or for concealing mistakes—it is hard to set any very clear or definitive standards or rules of thumb.

One almost has to come back always to the character, the integrity, of the individuals concerned. If he or she is worthy of trust, his judgment must be trusted as to when, and under what circumstances, confidentiality is required.

Unquestionably, however, certain individuals are by nature so obsessed with secrecy and concealment one suspects that, as infants, they were given to hiding their feces from their parents. One thinks immediately of Nixon. His former speechwriter, William Safire, reveals in his book, *After the Fall,* that Nixon was so secretive that, prior to his election, he mistrusted even the Secret Service men guarding him. His foreign-policy adviser, Richard Allen, wanted to bring him together with Anna Chennault, widow of the Flying Tiger general, who was pulling strings to block a Lyndon Johnson bombing pause in North Vietnam. "Meeting would have to be absolute top secret," wrote Allen, to "DC" (Nixon's "code" name). Secretive old "DC" scribbled opposite this reference to "top secret": "Should be but I don't see how—with the S.S. [Secret Service.] If it can be [secret] RN would like to see—if not—could Allen see for RN?"

Note that, for extra secrecy, he even writes of himself in the third person. "DC," even to himself, is RN.

We all know where this excessive passion for secrecy led. Kissinger not only had Safire's phone tapped, but even recorded—without their knowledge—conversations with such co-equals as budget director George Shultz. Writes Safire: "This tolerance of eavesdropping was the first step down the Watergate road. It led to eavesdropping by the plumbers, to attempted eavesdropping on the Democratic National Committee, and to the ultimately maniacal eavesdropping by the President, on the President, for the President, completing the circle and ensuring retribution. Eavesdropping to protect Presidential confidentiality led to the greatest hemorrhage of confidentiality in American history, and to the ruination of many good men."

Indeed, I sometimes think it is such *needless* passion for secrecy in many of our institutions, corporate as well as governmental, that has set off the present demand to wash, as it were, all public information in Macy's window. It has set off, as well, the unprecedented epidemic of public litigiousness, where every leader of any institution now has to consult his lawyer about even the most trivial decisions (I am currently involved in so many lawsuits my mother now calls me "my son the defendant").

So even while I defend the need for confidentiality, I argue for the utmost possible openness—for "leveling"—in every institutional hierarchy. In the 1960s, when I made some organizational studies for the State Department, I quickly learned that junior Foreign Service officers often decided not to tell their bosses what they knew from the field situation (because they believed the bosses would not accept it), only to learn later that the bosses felt the same way *but in turn kept silent* for fear *their* bosses would disapprove.

This went on, up and down the line, to the very top. Each privately knowing what was right, all enclosed themselves in a pluralistic ignorance, much like the husband who doesn't want to go to a movie but thinks his wife does, and whose wife doesn't want to go but thinks *he* does, so that both go though neither wants to. . . .

People in power have to work very hard to get their own key people to tell them what they do know, and what they truly feel. But the whole Vietnam mess is a succession of the failure, by people who knew better, to say what they really knew—either while in power, or after resigning because they could no longer stomach the ascending pyramid of lies and deceptions.

This leaves us with a paradox. The more we can establish *internal* truth—true openness, true candor, true leveling—within an organization and its hierarchy, the better able it will be to define, and defend, the proper areas of *external* confidentiality. . . .

Nevertheless, the national mania for "full information" is very much with us, and is now part of the turbulent social environment that every administrator must deal with. Dealing with it wisely will challenge all his tolerance for ambiguity. Freud's definition of maturity was the ability to accept and deal with ambiguity.

Among colleges, one result is already clear. The Buckley Amendment is laudable in its intent. But henceforth school and college administrators are going to be chary of putting any very substantive information into any student's record. What will wind up there will be so bland and general as to be useless (for example) to college entrance officials in making a considered judgment of an applicant's overall merits. If, for example, he had threatened to cut a teacher's throat but had not done so, he could scarcely be described as "possibly unstable"; the student or his parents might sue.

Edward Levi, the new Attorney General who was the dean of Chicago's law school and president of the university, is able to see these problems from all those perspectives. As a respected civil libertarian, he has publicly exposed flagrant abuses by the F.B.I.'s late director J. Edgar Hoover—most notably an asinine "Cointel" game of sending anonymous letters to both Mafia and Communist leaders with the intent to stir up conflict between them. At the same time Attorney General Levi has stressed the necessity of confidentiality, not only for government but private groups and citizens. As for Wilson's famed "open covenants," Levi quotes Lord Joseph Devlin: "What Wilson meant to say was that international agreements should be published; he did not mean that they should be negotiated in public."

In government, the Macy's window syndrome is going to make for greater inefficiency, because officials are going to spend more and more of their time processing requests for documents on *past* actions instead of applying the same energy to *future* actions. Levi points out that the F.B.I., which received 447 "freedom-of-information" requests in *all* of 1974, this year received

483 requests *in March alone.* "As of March 31, compliance with outstanding requests would require disclosure of more than 765,000 pages from Bureau files."

Such demands can, it seems, be self-defeating. One suit to compel disclosure of Secretary Kissinger's off-record briefing on the 1974 Vladivostok nuclear-arms negotiations yielded 57 pages of transcript, but three pages were withheld on grounds that "attribution to Mr. Kissinger could damage national security." More important, it raised the question of whether any future briefings would be equally informative—or, in some cases, discontinued entirely. As the Supreme Court observed, even while denying President Nixon's right to withhold the crucial Watergate tapes: "Human experience teaches that those who expect public dissemination of their remarks may well temper candor with a concern for appearances and for their own interests to the detriment of the decision-making process."

In the case of meetings of public bodies—school boards, college regents, and the like—the disclosure mania will make for more and more cliques which meet privately beforehand to agree on concerted actions subsequently revealed only at the "public" meeting. What is likely to emerge are the "pre-meeting-meetings" novelist Shepherd Mead described in ad agency conferences in his *The Great Ball of Wax.*

In every important decision likely to impinge on this new "right to know," there will likely be far fewer written, recorded discussions, far more private, oral discussions, far more tacit rather than "official" decisions. More winks than signatures ("Don't write. Send word.") if for no other reason than to avoid some new capricious lawsuit.

The public will be learning more and more about things of less and less importance. It will be poorer served by administrators trying to fight their way through irrelevant demands for "full information" about old business, to the neglect of attending to new business.

I am not saying that individuals who have been unjustly accused should not be able, as the Freedom of Information Act provides, to examine their own dossiers. Nor am I saying it is unwholesome for any government or public agency to be prodded out of its passion for hiding its mistakes under "classified" labels. That kind of file cleaning and purging is needed. Furthermore, scholars are finding the law a great boon in gaining quicker access to needed documents.

What I *am* saying is that in the long run we are likely to get better government, better decisions, by focusing our energies on finding leaders whose innate integrity, honesty, and openness will make it unnecessary for us, later on, to sue them or ransack their files. Edward Levi, it seems to me, cuts to the heart of the dilemma:

"A right of complete confidentiality in government could not only produce a dangerous public ignorance but also destroy the basic representative function of government. But a duty of complete disclosure would render impossible the effective operation of government."

15

Corporate Boards

Because of the radical changes in organizational life in recent years, corporate boards have become more important than ever. The old boys' club of the past, which did little more than rubber-stamp executive decisions, has given way to a diversified body that has both a legal and a moral obligation to scrutinize management and even to assume a leadership role, as the board of General Motors did recently.

The Crisis of Corporate Boards

But I'd shut my eyes in the sentry-box,
So I didn't see nothin' wrong.
                    —*Rudyard Kipling*

The now famous McCloy report, undertaken as a result of corporate actions ranging in their seriousness from "ethical insensitivity" to criminally liable behavior, concludes: "It is hard to escape the conclusion that a sort of 'shut-eye sentry' attitude prevailed upon the part of both the responsible corporate officials and the recipients as well as on the part of those charged with enforcement responsibilities."

What *is* the proper role of a board of directors in the conduct of corporate affairs?

A former Penn Central director admitted shortly after the fall of that corporation, "I don't think anybody was aware that it was that close to collapse." And Gulf's directors were clearly embarrassed by their company's illicit payments and other criminal actions.

The lesson from the Gulf and Penn Central situations is simple: unless directors have the right information and know how to ask the right questions, they tend to see only what they are told to look at; when something "wrong" happens they are apt to be jolted out of their lethargy by unfriendly lawyers rapping loudly on their sentry box.

The spectacular embarrassments of boards of directors in recent years have been warning flares lighting up the corporate skies. But fireworks unfortunately block from view the deeper, more difficult issues. At the same time, they may tend to act as tranquilizers, falsely lulling the sentry into confidence that the criminals have been purged and further liability curbed.

These unfortunate incidents have led to a stream of discussion, some of it shrill and not altogether rational. The business community, most of all, has been reexamining the composition, operations, and procedures of boards of directors. While there are indications that procedures and operations have been tightened, the increased public expectations about corporate performance are a long and nervous distance from realization.

It is still the case that most board members are, in fact, willing dupes of management. They are expensive, impotent, and often frustrated rubber stamps. They are subject to major litigation and are selected solely by their subordinates. They seldom understand their function, because of lack of proper orientation or education, and therefore they often meddle in management affairs. Old-timers who have learned the function of a board find that they have no way to assume those functions, given the history, tradition, or make-up of the particular board on which they serve.

An example of board confusion and the consequent resistance to change can be found in the case of Arthur Goldberg, a former member of the Supreme Court and Ambassador to the United Nations, who resigned as a director of T.W.A. on October 18, 1976, in a dispute over the directors' proper role. Goldberg attempted to establish an independent committee of outside directors to review the actions and recommendations of management. He argued that this group should be allowed to meet independently, without interference from "inside" directors, officials, or administrators, and should benefit from an independent staff of technical specialists. His proposal was turned down by the board, and Goldberg resigned.

In the two years since then, three dozen suits have been filed by disgruntled shareholders of the Penn Central Transportation Co. against the directors who had served that company prior to its receivership. These suits are based on a law imposing ultimate legal responsibility for the management of a corporate enterprise on boards of directors. More recently, the Securities and Exchange Commission has rebuked two outside directors of Sterling Homex Corp. for allegedly failing to obtain "a sufficiently firm grasp of the administrative, organization, and financial practices of the firm," accepting, according to the S.E.C. report, "superficial answers to questions put to management."

Players in a Chinese Baseball Game

The fundamental point here is that a gap in the corporate directors system exists between what state statutes say are the responsibilities of boards of directors and the realities of any board's operations. Furthermore, these statutes vary from state to state and from country to country and are in constant flux, creating for all public corporations a situation on the order of Chinese baseball—a game in which players can move the bases anywhere they want to.

The effectiveness of boards of directors is impaired by at least eight specific predicaments which are at once the source and the manifestations of the gap between a board's legal responsibility and its actual operations.

• *Board Composition.* Though there is now a mix of "inside" and "outside" directors on most boards, charges of "clubbiness" still echo. Too often, people join boards because of the prestige it gives them or the stipend it pays, or because it provides an ideal activity for retired executives who want to keep busy but who understand they will exercise no control. Outside directors, chosen by the president or a few inside officers, become "angel's advocates" of top administration. On the other hand, diversity on a board may lead to a "devil's advocate" stance in questioning and often leads to a "Noah's ark" syndrome: filing into the board room one-by-one must be representatives of every type—one woman, one black, one Jew, one consumer—whose appointments are supposed to absolve everyone of guilt for years of neglect.

Knowledgeable, experienced women and minorities do have a place on boards, of course, but to pick them solely on the basis of prior neglect is an insult; and it is likely to lead to a board made up of people who know little about the enterprise and who raise so many single voices and write so many minority reports that conflicts cannot be resolved and progress is blocked.

The predicament, then, is how to create a board responsible to the shareholders without creating a political Noah's ark of dissident voices.

• *The Increasing Role of Law.* A board of directors cannot operate with a corporate management unless there is mutual trust and confidence. But today the law's increasing influence tends to erode this. Board members are now so vulnerable to and skittish about expensive lawsuits, for example, that they can no longer rely on the word of the chief executive officer in making decisions. They must have everything in writing, duly notarized, which often means that a corporation's legal counsel and chief financial officer play a more important role than the chief executive officer. Moreover, the law—which has difficulty distinguishing between deliberate wrongdoing and an honest mistake —punishes only acts of commission, not those of omission. Under these conditions, almost any board will encourage management to do nothing rather than introduce risky innovation. And innovation—the willingness to take risks in the marketplace—has been the genius of American enterprise.

• *Accountability.* At what precise location does the buck stop? Why should a chief executive officer be fired for a wrongdoing unless the entire board goes with him? Isn't the board equally culpable? How can board members be the ultimate arbiters, the pinnacle of corporate power, without being accountable as well? The demarcation lines of power, responsibility, and culpability are vague and subject to many different interpretations.

- *Diminished Executive Responsibility.* Every reform resulting from infractions of the law or from criminal behavior on the part of corporations has diminished the responsibility of the chief executive officer. Examples: a company's financial officer is required to report to an auditing committee; nominating committees are to be composed of outside directors reporting directly to the board; the required standing committees on corporate social responsibility are to report directly to the board, rather than through the chief executive officer. Requirements such as these have the effect of preventing the leader from leading.

- *The Double-Bind.* The chief executive officer's responsibility, on the one hand, is to maximize the return on the shareholders' investment. But the competitive practices by which this obligation is fulfilled—normally quite within the accepted norms of the enterprise—may now suddenly be made illegal by a law that is in flux. To the traditional conflict between risk-taking and corporate security there is thus now added subtle pressure of the chief executive officer to assure his personal immunity to legal action.

- *Conflict of Interest.* What is the position of a member of a bank board, for example, who serves also as financial adviser for an enterprise which is a customer of the bank, or of a lawyer on a university's board who represents a firm that does business with the university? Such relationships deserve careful scrutiny. But it is difficult to find knowledgeable, sophisticated candidates for boards of directors who have no ties whatever to other institutions. Indeed, if all board members with multiple obligations were forced to resign, we would end up bringing people with no knowledge, sophistication, or advisory capacity into positions of power.

- *The Ambiguity of State Statutes.* There is no uniformity in the state statutes under which corporations are chartered, and many are highly ambiguous. For example, no state statute distinguishes between outside and inside directors. This general ambiguity frustrates attempts to define precise roles for directors, and as a result it is virtually impossible to answer the question, "Who's in charge?"

- *The Education of Board Members.* It is not possible for board members to ask discerning questions, to understand their full legal and managerial prerogatives and responsibilities, or to fulfill their responsibilities without orientation and education; but most begin their terms completely naive about their roles. There is little or no orientation. New board members usually receive at best a glossy confection supposed to provide a description of the institution, but which in reality offers little more than the typical annual report. Given the many competing pressures for their time and attention, how can board members learn what they need to know about the enterprise they serve and the environment in which it operates?

Confusion Rampant

There was a time when directors rarely had to dirty their hands or minds with anything but vague fiduciary responsibilities or the occasional selection of a president; today they find themselves surrounded by shrieking shareholders, lawsuits, and illicit practices, and they're confused. The major problem is the gap, more and more obvious, between their governing role as decreed by law and the reality of our complex, rapidly changing contemporary scene.

Rx for Corporate Boards

Not since the early days of the New Deal has the field of corporate governance been so astir with proposals for reform. Everybody's in the act, it seems—not just Ralph Nader, who nags, hassles, and litigates on behalf of corporate social responsibility or the ubiquitous Lewis D. Gilbert, who presses for reform and sues corporations to make them more responsible to the shareholder. There are also such substantial, temperate people as the participants in the 52nd American Assembly (April, 1977), whose conclusions on the "ethics of corporate governance" included a criticism that boards are "remote, insensitive, and not adequately reflective of the many publics they serve."

As a result of this growing clamor, two reforms are apparently being given serious consideration in Congress, state legislatures, and regulatory bodies:

- Federal chartering of corporations.
- Mandating certain proportions of public, independent (outside), and "special interest" directors, the latter including members of minority groups, to open up membership to broader constituencies and weaken the "clubhouse" atmosphere of board rooms.

That's all well and good. But I submit that these changes do not respond to the fundamental issues boards will have to confront in the coming years. These can be reduced to three:

- Restoring trust in the corporation. This can happen only if boards eliminate conflicts of interest; if board members make sure that the corporation they serve as the final repository of trust obeys the law; if they learn how to ask the right questions; and if they speak out on issues of public concern.
- Developing explicit guidelines regarding board and administrative accountability.
- Recruiting colleagues earnestly and well, making certain that the requisite combination of talents is present, and then providing a first-rate orientation program for new members as well as a plan for the recurring education of current members.

All the federal guidelines in the world will not improve the operations of boards until these rudimentary criteria are satisfied. Even if directors possess superior qualifications, no substantial change will occur unless they learn to ask discerning questions and to recognize and demand responsive, substantive answers. "Communication" or "dialogue" are not enough for dealing with important policy matters. Those require that directors ask questions, crystalize their views, and assert their informed opinions.

More Regulation, More Transience

Consider J. Pfeffer's recent analysis of the three basic levels present in all organizations: the *technical* level assures the organization's capacity to produce some item or service; the *management* level coordinates and supervises the technical level; and the *insti-*

tutional level assures the organization's legitimacy, credibility, and success in coping with its environment. Most companies are now well equipped in technology and management. More and more, a firm's success will depend on its ability and sophistication at the institutional level.

That is because today's organizations are open systems, operating in an uncertain and unforecastable environment. We only know that there will be more regulations, more controls, more personnel movement from one organization to another, more joint ventures, more mergers, more independence. Corporate success will depend increasingly on being able to understand the political landscape, to deal with bureaucrats, to promote the right laws and regulations, to mobilize public opinion, and especially to know the strategy of all these. Retail price maintenance, tariff protection, and licensing to restrict entry into a field are prime examples of issues now entering the institutional environment.

Management is concerned with the technical and managerial levels, and most boards are drawn unwittingly into this same arena. But that is wrong. Only where the board plays a pivotal role in institutional management by helping management understand and meet its new environment can organizations succeed.

Instead of being drawn into arguments about cost effectiveness, the board should be concerned about changing markets so that they may see cost effectiveness in a new dimension. They should have ideas about new markets, new technologies, new hopes. They should understand publications and public affairs, realizing that there are many new constituencies and patronage structures out there upon which the organization can founder.

My concern with the effective operations of corporate boards reflects a broader concern which I share with Harold Williams, chairman of the Securities and Exchange Commission: "The issue [his words] of the very legitimacy of the corporation itself." A more specific expression of this concern came from Assistant U.S. Attorney General Robert H. Morse upon the conviction of a supermarket president on criminal charges involving rodent infestation in one of the company's sixteen warehouses: "Only through incarceration will the business community be dissuaded from such conduct." The fact that the president had been assured by a subordinate that the violation had been corrected carried no weight in court.

Trading in the Marketplace of Ideas

How are we to re-establish the public's confidence in those who legally shoulder the responsibility for corporate America—the directors and top officials of the firm?

The most thoughtful and far-sighted chief executive officers and board directors are already striking out in new directions. Many of our top business leaders, men such as John deButts of A. T. and T., Reginald Jones of G.E., and Irving Shapiro of du Pont have long understood the role of political and social factors as major forces affecting business. Now they and many of their colleagues are spending ever-increasing amounts of time dealing with public affairs. And they do not take these responsibilities casually, for they realize that the marketplace of ideas, where business has not in the past fared especially well, may be as important as the marketplace of commodities.

Unfortunately, however, too many top managements remain oblivious to—or even resentful of—the new ball game. Their ambivalence leads to policies and actions which are almost always too late and too little, a "muddling through" process: a Washington office, a high-priced speech writer, exhausting public appearance schedules for top executives, and a big sigh of relief that, now, the problem is taken care of.

These "remedies" barely touch the surface of the contemporary realities facing our corporations; they are nothing more than anxiety-reducing actions taken without understanding the scope of the problem. Perhaps no one understands this better than du Pont's Irving Shapiro, who has written in the *Wall Street Journal:*

> A new breed of manager is coming to the fore, no less competent on the basics of business, let us hope, but much more of a public person. This reflects the fact that big corporations have become quasipublic in nature. You see this in the increased regulations of business, in journalistic probings of the corporate interior, in labor relations and in other ways.

Today you don't hear executives asking their trade industry associates to handle all of the relationships with Washington. What you see, increasingly, is company leaders doing the rounds in person, testifying, talking to congressmen in their offices, meeting with the regulatory agencies, meeting in the executive offices with people from the president on down.

. . . This process is healthy for the political process, as well as business. The business leader learns that you don't reach decisions in government the way an engineer chooses the most efficient process for producing a product.

An overwhelming majority of people in positions of responsibility in this country, whatever their walk of life, believes that we have a system that in practice works pretty well, and in principle is even better. What I see now within the ranks of business is a growing sensitivity, a growing sense of the possible and a growing set of talents to make the system work better.

16

Information Overload Anxiety (and how to overcome it)

Bureaucracies do one thing very well: create a paper blizzard. As a result, everybody I know has a guilt shelf, a spot, usually close to the bed, where unread memos, important mail, and other documents to be dealt with accumulate, along with yellowing back issues of the New York Times.

Little of what we insist on saving is of real importance. Most of it goes in one eye and out the other. This is a one-minute guide to dismantling your guilt shelf—without actually canceling your subscription to the New Yorker. *The piece was originally published in 1979, and the fact that a couple of the jokes are now unfathomable to me reinforces the central thesis that less is more, publications-wise.*

Lance Shaw uses a timer. The timer technique is invaluable for *enforced skimming* of the many magazines he must read. He gives himself *no more than 15 minutes* on the timer for each periodical. When it rings, he tosses out the magazine and moves on.

—EXECU-TIME

As we all know, ours is a poly-saturated, paper-polluted society; one in which power resides in those who have the right information. Fortunes are made or lost, careers secured or shot, health enhanced or damaged by one factor—information. In our society information equals power.

Coterminous with the information=power equation is a dramatic increase in a national nervous disorder: Information Overload Anxiety (IOA). This disorder is characterized by an obsessive-compulsive tendency to read everything about everything from anemones to zyzomys.* This is a killer disease made more pernicious by our collective and willful neglect of it. IOA is simple to understand: When the amount of reading matter ingested exceeds the amount of energy available for digestion, the surplus accumulates and is converted by stress and overstimulation into the unhealthy state known as IOA.

The so-called cures of this disease are as pervasive as the information-pushers themselves. The Lance Shaw bell-timer method is only one of the many regimens available at your local newsstand. Another comes from Nobel Laureate Herbert Simon speaking in *People* magazine: "Reading daily newspapers is one of the least cost-efficient things you can do. . . . Read *The World Almanac* once a year. What's happening you'll hear by lunchtime anyway." Russell Baker copes with IOA by reading only the obituary page to make sure he hasn't died, and, after ascertaining that, goes out to celebrate. Cutting back to Simon's starvation diet, or trading cerebration for celebration are only a few of the wonder cures. Many IOA victims have tried others: keeping the abstract and digest pushers prosperous with subscriptions; sitting through speed reading courses; having an assistant predigest industry news and then picking his brains. Some have tried newspaper fasting by shifting to liquid Cronkite, and other sufferers have stopped their intake cold turkey.

None of this is necessary. Neither must you be sentenced to a death row of sentences. As a reformed IOA patient, I can recommend a formula for dealing with this disorder. It is the surest, safest way to attain your basic minimum information requirement. For those of you who genuinely *want* to change and who feel that obituaries are unhealthy and the timer system mechanical, I present here the Fat-Free Daily Reading Diet guaranteed to satiate everyone from the freaky faster to the junk news junkie.

*The surgeon general has determined that reading further may endanger your health.

But first a word about the Fat-Free Daily Reading Diet (FFDRD). Who needs it? The ambulatory generalist, the well-educated, intellectually curious, overstimulated crisis expediter. In short, *you*. Second, many nutritious items beyond the ordinary nosher's price range, or inaccessible to those not adjacent to the kiosk on Harvard Square or the newsstand in the Citicorp Building, have been eliminated. Third, faddists need read no further. You will not again see mention of such exotica as *Cahiers du Cinéma, Blood, Iron Age, Running Times* or *Hungarian Art Nouveau*. These are to be put immediately into your Cuisinart and pulverized into fruit smoothies. I choose my products on the basis of their health-giving balance, practicality, consistency, and absence of bulk. Less is more, so to speak. This plan is based on experiential boredom, and has been laboratory tested by patients who once seemed hopelessly mired in decadent reading habits, but who are now reading normal healthy lives.

Basic Maintenance Plan

FFDRD is the reading plan to follow the rest of your life. Once comfortable with it and sure of your new control, you may want to modify it to include in your daily diet the *Boston Globe*. But only as a substitute for a news magazine. And remember, only *every other* page of a daily is allowed—and first throw out the Metro and Entertainment sections.

Group I: Newspapers. One daily. But proceed slowly, and check first with your physician. Choose from the following: the *Wall Street Journal*, the *New York Times*, the *Washington Post*, or the *L.A. Times*. Absence of bulk clearly makes the *Wall Street Journal* the preferred choice, and none of the other three is delivered outside its region on the date of publication. Nor does the *WSJ* suffer any episodic regional virus, such as the *Post's* Potomac virus, leading to feverish stories about the identity of Jimmy Carter's tennis partner at 3:05 yesterday, or the *NYT*'s Big Apple syndrome, leading to information about where one can buy Albanian sausage on Second Ave. or on which Thursday you can dance the hora on Prince Street with your children.

The *WSJ*'s editorials are peppy, nutritious, and infused with vital life-giving substances. As a daily regimen you will find its reviews of cultural events hearty and fat-free.

Group II: News magazines. One to be selected from the following: *Newsweek, Time, U.S. News and World Report.* Diet must be limited to one. Ingesting two reveals serious addiction. Since the basic contents of *Newsweek* and *Time* do not vary significantly, reading both will throw your metabolism out of kilter. The *U.S. News* has slimmed down considerably and has at least one fat-free item each issue, but *Time*'s essays on Masters and Johnson, and the post-Vietnam American consciousness are superb supplements to a restricted diet. *Newsweek,* however, is less caloric. In five out of eight tests, *Newsweek* was found to contain fewer carbohydrates, and it regularly contains such tonics as Meg Greenfield and George Will.

My preference for both *Newsweek* and *WSJ* is also based on the belief that a healthy agent locates pain-in-hiding and identifies potential problems in the soft tissues of a listless society; it recognizes symptoms and points them out: racism *before* Dr. King's march, pollution *before* Rachel Carson, sexism *before* Betty Friedan, bureaucratization *before* William H. Whyte, Jr., consumer protection *before* Ralph Nader.

Group III: General Culture and Ideas. Few temptations here. Needed is one unsweetened, streamlined packet that reviews books, movies, dance, theater, music, TV, the plastic and performing arts, architecture, urban aesthetics, graphic design and the commercial arts. No such item exists in the U.S. In waiting rooms try *Vogue;* in England sample *The Listener;* on Sunday read the *Times.*

A warning about the Sunday *Times:* It has unquestionably brought more patients to terminal IOA than any other single diet component. It must be read only under doctor's supervision, and you must first remove bulk—information on street dances in Soho, all lingerie ads, travel pages, reports of recent polo deaths in Rhodesia. Leave in essential ingredients: discussions of architecture, drama reviews, essays on arts and culture. Again, be forewarned: Forty-year-old John C., a former patient, had cut his information overload to a tolerable level. At a neighbor's house he picked up a copy of the Sunday *Times* and read it voraciously. Next we heard he was into *Saturday Review, New York Review of Books,* and when last seen was munching a forbidden *Harper's* and stuffing heavy paper items underneath his sweater. Remember: Reading the Sunday *Times* is like eating one peanut.

Group IV: Reference Books. Do not be hooked by book clubs or by friends' seductive shelves of *Encyclopaedia Britannica.* The typical IOA patient may start off with a pocket dictionary, then cautiously proceed to the rather disappointing *London Times Atlas.* By then there's no stopping him. The two-volume *Who's Who,* the addictive *Oxford English* whatever, a dictionary of quotations, of phrase origins, of slang, of golf. From that it's on to almanacs: farmer's, gardener's, jogger's, swinger's, ball club owner's and—there you are, off the wagon again.

This diet includes only one reference book—a dictionary with intellectual heft and clarity; an untrendy dictionary, but one that doesn't shy away from "DNA" or "glitch"; a dictionary with *helpful* pictures, but no drawings of nailheads or cows; one with a first-rate etymology, and with type that can be read without an accompanying magnifying glass which makes you, no matter how hardy, feel like a geriatric case. The ideal dictionary also includes a section of the most commonly used German, French, and Spanish words, a basic manual of style, and some major dates in history. The dictionary that comes closest to meeting our minimum daily requirement is the *Random House Dictionary of the English Language*—unabridged. But unlike the new you, it is so overloaded with information you must not attempt to carry it about with you; instead, leave it in an accessible place with good light.

Reference supplements: To help you adjust without regressing to book clubs, libraries, and binges resulting in a three-year subscription to *Mother Jones* or *Popular Mechanics,* I refer you to the following extras containing no fat, but which may be read only by those who have their appetites under control.

The National Directory of Addresses and Telephone Numbers, a Bantam paperback, includes an index of 50,000 useful addresses and phone numbers: all U.S. companies with annual sales of $10 million and over, governmental agencies, foundations, educational institutions, museums, ballparks, symphony orchestras, public libraries, hotels, airlines, hospitals and all those organizations you've been wanting to write hate letters to all these years.

Free Items: The Morgan Guaranty Bank in New York offers a monthly report on the U.S. economy, and Metropolitan Life publishes a monthly *Statistical Bulletin* tracing important demographic trends and their consequences. Good, but not quite free, are *Manas,* an idiosyncratic charmer tending toward philosophy and history of ideas, and *Brain/Mind,* a model of less-is-more with a neat four-page format containing almost as much information as one issue of *Psychology Today.*

The Basic Harold Laski Reading Diet

People vary in their reading metabolism. Some, like former president John F. Kennedy, have a high IOA threshold. It is rumored that he devoured everything in sight and still remained free from nagging symptoms. Harold Laski was famous for this too. He could digest an average book in minutes with only a bare expenditure of energy. For the lucky few who do not require a strict regimen, the following may be included as a supplement to the Basic Maintenance Program.

Group V: Management and Business. We live in a corporate society—the daily press's best-kept secret. If you can find the business section, you'll see it stuck somewhere between sports and classifieds. Business/management magazines, however, are burgeoning and old standbys with expensive facelifts like *Fortune* no longer resemble thick racing forms. Preferred by dieters: *Fortune, Barron's, Forbes* and *Business Week.* If restricted to these, the prudent patient should have no weighty problems. A favorite is *Business Week,* which is timely and easily digested. It tends to items you can take in from beginning to end on one page. Its longer features are superb fare: "Washington Outlook," "Social Issues," and "Book Reviews."

Group VI: Science and Technology. No noshing here. This group contains no junk additives. The "Comment" section in *Sci-*

ence, for example, is a heavy meal-in-one—substantive, and comprehensive. If you're into "sieve areas in fossils" or "food-webs and niche space," *Science* just may be your main fare. *Scientific American* is well-known, but too often gives the specialist too little and the non-specialist too much. It leaves you either hungry or overstuffed. Its design, however, is mouth-watering, and its regular features offer thought for food.

The New Scientist comes from England and is staffed with perhaps the most able science-writers around. It is often intelligent and timely, so don't be put off by items like "the courtship patterns of the red-vented bulbul." It gives the slimmed-down layman an overview of everything from the economics of micro-processors to fossilized footprints.

If there were one all-purpose compound which included the most basically sound ingredients of all the science magazines, I would happily recommend it. And I hear by the grapevine that the publishers of *Science* will soon come out with something like that.

So there you have it. The Fat-Free Daily Reading Diet. If you stick with it, you will enjoy the new trim-reading-you—the emaciated intellectual inside you just waiting to emerge. And if you fail? Well, there are weekly meetings of Readers Anonymous in every major city. Consult your National Directory—and join me.

IOA Self-Evaluation Test

Directions

Please try to answer all of the following questions. If you are uncertain, go on to the next question and come back to the one(s) you left incomplete. Before starting test, please take your pulse per 10-second interval. If your pulse-rate rises in responding to 15 or more of the questions, call *your doctor immediately. If your pulse-rate increases in responding to 10 questions, do not continue. We do recommend, however, in such cases that you do make an appointment with your physician.*

Start

DOES YOUR PULSE RISE WHEN YOU:

1. Don't know if the ZAP count of a microwave oven is more or less than 4 dental X-rays or a Sunday drive within a 150-mile radius of Harrisburg?
2. Wonder if you should invite a quark for cocktails next week, especially one of the "charmed" quarks?
3. Put on your carcinogenic, flameproof 'jammies or your non-carcinogenic 3 A.M. flashpoint "nighties?"
4. Debate insurance against crowbars falling from space platforms or emergent Loch Ness monsters?
5. Calculate your chances of myocardial infarction from family history, exercise habits, stress, cholesterol in your bile (alleviated by bran), lipids (reduced by yogurt), smoking, weight, or all of the above?
6. Read conflicting reports about the overall value of jogging?
7. Argue about whether we will have one more year of double-digit inflation and Fred Kahn? About whether we will have one more year of double-digit inflation and James Schlesinger?
8. Wonder if Black Holes and Pot Holes have nothing/something/everything in common?
9. Question whether the Supreme Court is threatening or protecting the rights guaranteed under the Fourth and First Amendments? And are not completely certain what those rights are (or were)?

Please Take Your Pulse

10. Can't be sure if dolphins are really getting admitted to Harvard Medical School?
11. Can't describe in 5,000 words or less President Carter's energy policy and whether or not it will make any difference (Carter's energy policy, that is)?
12. Can't describe in 3,000 words or less what SALT II is and whether or not it will make any difference?
13. Aren't certain whether Pres. Carter is getting stronger or weaker, softer or harder and whether or not it will make any difference?
14. Can't tell whether Pres. Carter shifted his hair style from left to right or right to left and whether or not it will make any difference?

Please Take Your Pulse

15. Don't know how serious George Steinbrenner III is about re-hiring Billy Martin for the 1980 season? Or how serious Johnny Carson is about quitting NBC? Or how serious Jerry Brown is about Linda Ronstadt?
16. Cannot respond to a question asking you to identify the winners ('79) of the National Book Award and if they are worth buying (in hard cover)?

17. Can't for the life of you think of the *questions* to the following answers taken from the May 12, '79 *New York Times* quiz. The answers were: 1) Diet soda and hard liquor; 2) "Utatuba" (Hint: a large granite structure that once stood outside an East Side art gallery; 3) Bob Hope; 4) Does not; 5) Harold Clurman; 6) Philip Caldwell will succeed him.

18. Haven't heard of Sparky Anderson?

19. Are uncertain whether it's worth driving 500 miles out of your way to attend a performance of the Houston Opera? Are even more confused about 1,000 miles?

20. When you can identify without trouble the *last* names of the following: Lennie, Woody, Scotty, Maggy, Jimmy, Billy, Jerry, Ronnie, Reggie, Archie and Deng (Xiaoping), but are bewildered when it comes to knowing the *first* names of the following famous personages: Jacuzzi, Franco, Sadat, Marcos, Masters, Johnson, Garp, Hayakawa, Khomeini (Hint: It is *not* Ayatollah), Brezhnev, Giscard d'Estaing, Somoza, Tito, Gilbert, Sullivan, and Hildegarde (not Neff).

21. Don't know what the following five men have in common: Alan Ladd, Jr., Henry Ford II, Johnny Paycheck, Michel Bergerac, and "Catfish" Hunter?

22. Can't match the names (left column) with the countries they represent:

Yao Grunitsky	Togo
Ionatana Ionatana	Tuvalu
Ousman Ahmadou Sallah	Gambia
Aristides Royo	Panama
Abel Muzorewa	Rhodesia
Mamady Lamine Conde	Guinea

Not Too Far to Finish, Please Take Pulse

23. Have never heard of Denis Thatcher?

24. Don't know if what the following have in common are any of these: 1) Office or house furnishings; 2) Slang for cocaine; 3) A purloined list of the Dallas Cowboys' draft choices for the 1980 season; 4) Prime Minister Joe Clark's cabinet members; or 5) Baseball players. After reading the following names, please circle one: Speed, Reed, Dent, Harrah, Rice, Bonds, John, Nettles, Lamp, Bench, Rose, Lemon, North, Rivers, Fell, Valentine, Swisher, Spikes, Cage, Hood, Page, Office, Gross, Wise, Klutts, May, Abbott, Waits, Porter, and Ford.

Please Take Pulse and Compute Your IOA Score

17

Our Federalist Future: The Leadership Imperative

> *Today I have the same confidence in the inevitability of federalism that a quarter century ago I had in the triumph of democracy. James O'Toole, who holds the University Associates' chair in management at the University of Southern California, is the senior author of this essay.*

> The structure of the organization can then be symbolized by a man holding a large number of balloons in his hand. Each of the balloons has its own buoyancy and lift, and the man himself does not lord it over the balloons, but stands beneath them, yet holding all the strings firmly in his hand. Every balloon is not only an administrative but also an entrepreneurial unit.
>
> —*E. F. Schumacher*

The Geopolitical Solution: A Template for Corporations?

In these turbulent times prudent mapmakers work on Etch-a-Sketch pads. Political boundaries change almost weekly as new nations emerge with varying degrees of anguish from the disintegrating empires of yesterday. No part of the world is exempt. From the Balkans to the British Isles, from the banks of the St. Lawrence to Guangdong Province on China's muddy Pearl River, ethnic and linguistic groups are wrestling—often at the cost of their lives—with a fundamental challenge of our era: We'll call it the Iceland Dilemma. In a sentence, the horns of that dilemma are represented by the choice between the advantages of small-country autonomy, on the one side, and the benefits of big-country economies of scale, on the other.

Bleakly beautiful Iceland is being pulled and shaped by these two powerful but opposing forces. On one side is the Icelanders' fierce pride in their nation's Viking heritage. This pride has led the population of Iceland (in total, some 250,000 shivering souls) to form a committee to give Viking names to concepts that even their visionary national hero, Leif Ericksson, could never have imagined. Hence, in Iceland a computer screen is called a *skjar* (the ancient word for a "window" on a traditional turf house).

But that passion for what is uniquely theirs is only one side of modern Icelandic values. Even as Icelanders quote traditional sagas and support laws to require citizens to choose names for their children from an approved list of "pure" Icelandic origin, those same citizens are eager to enter into active participation in the global economy, to become a part of the highly competitive modern world of international technology, commerce, and finance.

Icelandic society is thus a vivid example of what philosophers once called the tension between the tribal and the universal. How to balance tradition with the desire for economic progress, how to be true to oneself while being a partner, and how to sing solo but be in the chorus at the same time—these are the essence of the Iceland Dilemma. Indeed, the entire world today is grappling with the need to strike a balance between nationalism and globalism. And that need is likely to grow more pressing as the new millennium unfolds, considering that there are more than 5,000 restless "nationalities" in the world but only 166 nation-states—so far. Clearly, the number of nations should be viewed as just penciled in, and can be expected to increase dramatically as countries divide and subdivide even further in coming years, all the while seeking simultaneously to be a part of the New Globalism.

Fortunately, there is a generic solution to the Iceland Dilemma: *federation.* Thus, many Icelanders would like to escape their own particular version of the dilemma through an exogamous marriage with the European Community (EC)—joining their fortunes to those of peoples who have little interest in the purity of the Icelandic tongue or the preservation of Icelandic culture but who can provide the political and economic clout Iceland needs to be a player in world markets. Indeed, the European Community is the surpassing model of the federalist future. With twelve member-states (and counting), the EC is now seen by some three score ethnic groups living in the twenty-five nations situated between Reykjavik and Riga as the best means for them to unify for overarching political and economic purposes, while at the same time maintaining their cultural integrity. As we shall see, not only is such confederation the most practical resolution of the Iceland Dilemma for nation-states, but it can be equally beneficial as a strategy for business corporations.

Government (and Corporate) Federalism

Unlike monolithic forms of government, federations are alliances of more or less independent states, often with little in common but their desire to share in the benefits of swimming in a larger pond. The most durable example of confederation is Switzerland, where a workable union of divergent cultures has survived for more than seven hundred years. In modern Switzerland there are twenty-six semiautonomous cantons (and half-cantons), which together compose four major cultural groups, each with its own language and customs. Perhaps the most convincing argument in favor of federalism is that the Confederaziun Helvetica endures despite this remarkable diversity (tolerating even the reactionary values of one half-canton that is the last political body in the Western world to deny women the vote in local elections). In general, federations allow member units to pursue their unique—even quirky—interests, to realize their distinctive possibilities, and to address their special needs, as long as the units do not compromise the rights of other members or the needs of the alliance as a whole.

In that most successful of federations—our own resilient alliance of states—the whole is greater than the sum of Alabama, Alaska, Arizona, and the other disparate but essential components of the Union. In these rapidly changing times, such federations as the United States work better than monolithic nations (like the former, misnamed, Union of Soviet Socialist Republics) because they offer flexibility as well as strength. By their nature federal systems recognize the legitimacy of alternatives, of more than one possible response to a given challenge. If a federation were a poem, it would be not the epic saga of a single national hero but something like Wallace Steven's "Thirteen Ways of Looking at a Blackbird."

Committed to a single vision and course of action, a unitary government is often too slow to respond to changing conditions. In contrast to the singular stance of the monolithic state, federations are nimble by nature, accustomed to considering a full repertoire of responses. While the unitary nation goes for all or nothing, federations multiply the options and reduce the risk. In theory, at least, federations are also less prone to the ethnic animosities that are the ugliest aspects of hyperpatriotism. The very existence of a federation is implicit recognition that there is strength in diversity. In homogeneous groups outsiders are too often seen as monsters, devils, or obstacles on the road to "racial purity." But it is much harder to dehumanize outsiders in a heterogeneous alliance in which others are viewed as peers and partners (albeit ethnic vilification is not impossible in federations, as the former Yugoslav republics sadly demonstrate). James Madison, the guiding light of American federalism, argued that the true virtue of such a system lies in its recognition of the natural tendency toward the pursuit of self-interest—and in its built-in mechanism to counter that tendency by protecting the rights of minorities from "the tyranny of the majority."[1]

Because federalism allows constituent units to maintain their integrity while unifying for common purposes, it is not surprising that the form is now a major trend in business as well as in government. For if "centralization is the deathblow of public freedom," as Disraeli said, it is equally the deathblow of corporate innovation. For that reason, many of the world's most influential business leaders are creating new kinds of corporate confederations with numerous semiautonomous units, often in far-flung countries, joined together only to allow them all to succeed better in an increasingly competitive global economy. Examples include Benetton, Coca-Cola, and the newly formed Asea Brown Boveri (ABB). These companies have become models for international orchestration, influencing the likes of General Electric's CEO Jack Welch. Welch is creating at GE a new corporate federalism which he describes as "boundaryless."

Significantly, the characteristics of successful national and corporate confederations are nearly identical. Moreover, the following characteristics of federalism have remained constant since they were first described by Madison in the late eighteenth century, and thus appear to possess almost universal validity.

- *Noncentralization.* In federations power resides in many semiautonomous constituent centers, deliberately diffused for the purpose of safeguarding the freedom and vitality of those units. This *non*centralization should not be confused with commonplace *de*centralization (typically characterized by an all-commanding central authority that unilaterally delegates specific, limited powers to its subordinate units). In sharp contrast, a true federal system is contractual and power cannot be rescinded unilaterally or arbitrarily by the central government (or central headquarters). For example, the corporate staff at one of America's most truly federalized corporations—Dayton-Hudson (DH)—cannot change the rules of the game that affect its Mervyn's and Target divisions. As with the Swiss confederation, such changes may occur only as the result of mutually respectful negotiations, a process that is prescribed in Dayton-Hudson's "constitution."

- *Negotiationalism.* In federations decisions are made in an ongoing process of bargaining between the units and the central authority—and often between the units themselves. Thus there is shared decision making, and the units have a guaranteed voice in defining their financial, administrative, and other obligations to the central body. This doesn't mean that Coca-Cola's distributors and bottlers dictate terms to CEO Roberto Goizueta; nor, as we shall see, does Goizueta dictate to them either. Rather, it means that terms and conditions are negotiated and contractual.

- *Constitutionalism.* In federations there is a written (occasionally, unwritten) covenant that binds the allegiance of the units to the basic purpose, mission, philosophy, and principles of the overarching institution. Often this constitution spells out the mutual rights and responsibilities of all parties. Constituent units in turn may be free to have their own constitutions, as long as these do not violate the basic principles of the articles of federation. Much like the United States has its Constitution, ABB has a 21-page "bible" that lays out the principles by which the company operates, and Dayton-Hudson's 118-page "Management Perspectives" serves much the same purpose.

- *Territoriality.* In federations there are distinct boundaries between the constituent units. In the case of nations, these geographic boundaries may be based on ethnicity or tradition. In corporations the boundaries can be based on business or product line. (With franchisors like Coca-Cola and Benetton, the boundaries are often geographic.)

- *Balance of Power.* Federations seek balance not only between the central authority and the units but also between the units (the nineteenth-century confederation of German states failed, in no small measure, because Prussia overdominated its weaker partners). Part of the negotiations that led to the 1980s merger of Dayton-Hudson and Mervyn's concerned the relative role the California chain would play in the established, Midwest-oriented pecking order of DH's other retail units.

- *Autonomy.* In a federation the units are free to experiment and be self-governing to the extent that they do not violate the fundamental principles necessary for the maintenance of the union. Of all the characteristics of federation, this is the most difficult to achieve and maintain. As students of the Civil War are aware, the American Union was nearly dissolved because of conflicting interpretations of this principle. Over far less morally significant matters, Benetton recently found itself sued by an angry franchisee who claimed that the corporation was imperiously dictating policies that ran counter to the spirit of the alliance. (We return to this important issue later).

The Necessity—and Fragility—of Federalism

Madison argued that these traits of federalism become necessities when an organization reaches a certain size. While the founders could imagine successful unitary republics on the scale of the Athenian city-state (or Renaissance Venice and Florence), they argued that even the original thirteen states were too big to function monistically. Their insight seems apposite to corporations as well. Small, well-managed companies like Ben and Jerry's, Herman Miller, and Chaparral Steel operate effectively within a unitary structure and culture and demonstrate little need for federalism. While Madison recognized that size alone is not the only relevant criterion for choosing federalism—diversity is another—it is clearly the single most important reason. Is it coincidental that almost all large social and economic institutions that find themselves in trouble today are unitary in form? From the People's Republic of China, to IBM, to the Los Angeles Unified School System, almost all such monolithic organizations could benefit from a heavy dose of federalism.

In this regard it is significant that many of history's most successful giant institutions—the Catholic church, the Roman Empire, the General Motors corporation, to cite three rather large examples—enjoyed their finest hours during periods when they were structured along roughly federal lines. For instance, GM reached its pinnacle in the late 1920s, when it briefly approached Alfred Sloan's original concept of six confederated divisions—and GM was never so *un*successful as it was in the late 1980s, when it had all but abandoned the last remnants of true divisionalization (even producing Buicks on Chevrolet assembly lines).

The GM example also illustrates the essential fragility of federalism, an inherent instability that stems from the aforementioned tension between (a) the needs of central authority to exert power and (b) the rights of the units to autonomy. The art of leadership in a federation is to preserve the balance between those ever-shifting forces. History shows how difficult that art is in practice. Like GM, most federations have a tendency—fatal in the long term—to overcentralize and homogenize. The old USSR is a classic political example of this pattern. And the root of the Soviet problem wasn't simply Communist dogma. Under Margaret Thatcher capitalist Great Britain also did not go far enough in the devolution of authority to the constituent parts of the United Kingdom (and the new democratic Russian "federation" seems to be regressing toward Soviet-style centralization). The former Union of South Africa was once a relatively effective (but undemocratic) federal state. Then, in the 1950s, power was centralized in order to impose apartheid on the reluctant English-speaking provinces. The result was the erosion of autonomy and the creation of a unitary (and even more undemocratic) republic.

At the other extreme, the United Arab Republic (a short-lived marriage of convenience between Syria and Egypt) had nearly none of the characteristics of successful federations listed earlier, and consequently crumbled as if constructed of Arabian sand. And conglomerate corporations—like Dart Enterprises in the 1970s—typically disintegrate (or degenerate into mere holding companies) when there is no unifying vision, constitution, or federal structure. As with so many conglomerates of the 1960–1970 era, the parts of Justin Dart's once mighty empire are now scattered across the Fortune 500.

The fundamental and continuing question facing all federations is this: What powers rightly belong with the central authority, and what powers should be reserved for the constituent units? Madison believed he had solved the question with the U.S. Constitution and Bill of Rights, which basically limited the power of Washington to matters of defense, foreign affairs, and regulation of interstate and international commerce. In theory he may have been on target, but in practice he failed to anticipate a slew of complex problems that could not be relegated simply to one box (federal responsibility) or the other (states' rights). The problem came to a head over a moral issue—slavery—and has recurred time and again, particularly in relation to other issues of human rights and, more recently, to environmental questions (for example, we are now agreed that no state has the right to pollute air that blows over neighboring territory). In fact, there has been a steady erosion of the power of the states, particularly in areas of fundamental moral principle. Until recently Americans had cherished the belief that setting educational policy was a state's right. Yet former President George Bush—an unapologetic opponent of centralization—argued that the education of the nation's youth is of such overarching importance that it cannot be left solely to the discretion of the states. Significantly, his solution to the problem was a classic example of federalist thinking: The national government will set performance standards, and the states and localities will be free to find the most effective ways and means of achieving those standards. This is a specific illustration of federalism's most basic playing rule: The central authority establishes the why and the what; the units are responsible for the how.

Significantly, it is this principle that has been violated, until recently, by almost every business corporation that has attempted to become a confederation. It was this "principle of coordination without losing the advantages of decentralization" that Alfred Sloan attempted, and failed, to define for GM in his classic 1921 "concept of the organization" study.[2] Some years later, in 1963, Sloan admitted he was "amused to see" that, in trying to simultaneously achieve coordination and specialization, his "language was contradictory." While Sloan never abandoned his wish to resolve his corporate version of the Iceland Dilemma, in practice there was a steady erosion of "states' rights" at General Motors almost from the day he unveiled his federalist structure for the corporation.

In fact, Sloan, his colleagues, and their many generations of successors were never comfortable with the leadership style required for federalism to work. The system requires several things of those in central authority: faith in the power of people to solve their problems locally; willingness to forgo the satisfaction of exercising command and control; and understanding that, in complex systems and turbulent times, no one individual or group possesses enough knowledge to manage the jobs of everyone else in the organization. Sloan—and tens of thousands of managers around the world who were to become his disciples by way of the business school gospel of "specialization/differentiation" cases—was never comfortable with such basic assumptions about organizations and leadership. Indeed, the most famous practitioner of going-through-the-motions federalism was Harold Geneen, who had the form of confederation down pat at IT&T but lacked the essential "feel" for the technology of collaboration to make the system function entrepreneurially.

Therefore, in spite of the rhetoric of decentralization, neither GM nor IT&T (nor the countless giant corporations modeled after them) was ever a true confederation. At least not until the unprecedented turbulence of the late 1980s began to force corporate executives to reinvent Mr. Madison's (and Mr. Sloan's) marvelous notion . . . this time with feeling.

Resolving the Big-Company versus-Small-Company Dilemma

Here's the circuitous path by which corporate America has finally arrived at federalism: Historically, America has been the land of the entrepreneur. In no other country have entrepreneurs been revered in legend the way they have been in the United States. Until midcentury the mythical Horatio Alger and the historical Henry Ford were genuine heroes (almost like Napoléon in France or Lenin in the Soviet Union). But by the end of World War II, the entrepreneur was an endangered species in this country. In the years immediately following the war, so-called organization men—the risk-averse children of the Great Depression—had little interest in chancy careers in the corporations they dominated. Surveying the structure of industry two decades after the war, the renowned Harvard economist John Kenneth Galbraith declared entrepreneurialism to be an anachronism and hailed the apotheosis of professional managerialism and giantism. "The planning system" (as he called the industrial form emerging in the 1960s) was to be dominated by a few monolithic corporations working in close concert with government ministries.[3] No longer would dozens of small firms compete within a given industry or for a given market. In Galbraith's brave new world, it would be USA, Inc. versus Japan, Inc. versus Germany,

Inc. (or, more specifically, General Motors versus Toyota versus Volkswagen).

In fact, Galbraith was almost proved right: In the 1960s and 1970s, the big did get bigger and the number of competitors was reduced. For example, in the jet engine industry there were just three giants: GE, Pratt and Whitney, and Rolls-Royce—the first two of which built the largest factories in the Western world in pursuit of the holy grail of economies of scale. Similarly, by 1970 most major U.S. industries were dominated by one or two mammoth firms: GM (autos), U.S. Steel (metals), IBM (computers), Exxon (oil), Bank of America (finance), and Sears (retailing). In Europe the pattern was even more pronounced: The Italian government gobbled up scores of small companies and conglomerated them into giant, state-owned groups; in Britain nearly the entire auto industry was amalgamated into one giant firm. This "New Industrial State" was the right way to go, according to Galbraith—and most Europeans believed him: Witness Jean-Jacques Servan Schreiber's *Le Défi Américain.*

Everyone knows what happened next: Within a decade GM had been badly embarrassed not only by smaller Ford and Chrysler but also by a passel of even smaller Japanese and German firms; U.S. Steel was being chopped up by minimills; IBM had literally hundreds of smaller competitors; Exxon's megalomania had led it to acquire a bushel of small, successful, high-tech companies—and then to micromanage them into failures; and both Bank of America and Sears were being niched to death by, respectively, financial boutiques and numerous small competitors in the retailing industry. Thus, by the mid-1990s the entrepreneur was not only back from the brink of extinction; he (and, now, she) was said to be in ascendancy. In the Reagan era the giant corporation seemed destined to the fate of the brontosaurus, and George Gilder was crowing (while Galbraith was eating crow).[4]

While there can be no doubt that the 1980s belonged to the entrepreneur, Mark Twain's oft-quoted line "News of my death has been greatly exaggerated" may be finding a parallel in the life cycle of large corporations. Today it seems wildly premature to join Gilder in assigning big business to the ash heap of history. This is not to defend the past behavior of the many complacent industrial giants who squandered America's precious assets in the 1960s and 1970s—their self-defeating human resources policies, suicidal customer relations, misguided planning, and faulty financial assumptions are beyond rational defense. Yet there is no evidence to suggest that the current denizens of the Fortune 500 are collectively about to go out of business—not next week, not next year, not in the next decade (and not even in the next century). There are several reasons that large corporations continue to survive—and it behooves the enthusiasts of small business to keep these in mind:

Some Inherent Advantages of Large Corporations

- They possess economies of scale in finance, purchasing, distribution, advertising, service, R&D (and, arguably, manufacturing).
- They are able to undertake *global* marketing.
- They have resources to protect themselves against cross-subsidization (dumping).
- They are able to maintain a large, diverse bank of skilled people (which allows them to invest in lengthy training for future assignments and to survive the loss of key individuals).
- They possess the organizational wherewithal and managerial know-how to bring more than one project at a time from the idea stage to full development.
- They provide key employees with a relatively high level of security and financial benefits.
- They are able to undertake the long-term planning and commitment of resources needed for giant, capital-intensive products (for example, a jet airplane).
- They have social clout with government and unions.
- They can afford to undertake basic research and to make slow, costly, incremental improvements in process technology.
- They have stability because they can afford to be integrated backward (to suppliers) and forward (to dealers).
- They tend to be diversified and hence less susceptible to va-

garies of the economic cycle (and less vulnerable if one or two key products fail).

While such stability, security, predictability, synergy, and discipline are at best theoretical advantages of large business, sufficient examples can be supplied to support most of these claims. After all, what small firm would not want to have the financial, service, marketing, distribution, purchasing, and R&D punch of an IBM? Especially—and this is the key point—if those benefits of size could come without the *dis*benefits of bureaucracy.

Which brings us conveniently to the advantages of small- and medium-size businesses. Because there are so many static mom-and-pop firms that cannot serve as models of eminence, we have in mind here the characteristics of the fastest-growing entrepreneurial businesses cited recently by *Inc.* magazine:

Some Inherent Advantages of Small Firms

- They tend to be lean, agile, dynamic, and flexible (nonbureaucratic).
- They are close to their customers and thus sensitive to (and fast to react to) shifts in market demand.
- They are run by managers who often are owners and are therefore highly motivated by their equity positions.
- From top to bottom, nearly everyone in the company has direct, ongoing personal knowledge of most aspects of the business.
- Their employees are motivated by the human scale of the organization, by peer pressure, and by knowledge of how their roles contribute to overall company performance.
- They have excellent upward, downward, and lateral communications.
- They attract the most creative, energetic and risk-taking individuals (indeed, there is a "brain drain" from large to small companies).
- They have a focused orientation on a single product or related line of products.
- They have short production runs and can thus customize products and keep a constant watch on quality.

The Big Mimic the Small

These impressive advantages are in fact the very characteristics of small firms that almost all large corporations today are attempting to capture through frantic attempts to alter their "corporate cultures." In order to "get close to customers," to "become people-oriented," and to "focus on quality," giant corporations around the world are experimenting with intrapreneuring, gain-sharing, team approaches, spin-offs, product-line focusing, specializing, downsizing, dis-integrating, subcontracting, and decentralizing—in effect, emulating what small companies do naturally.

Hence, in this paradoxical world we are faced with yet another fine irony: While entrepreneurs are trying to capture the advantages of large firms, managers of large corporations are at the same time attempting to behave like entrepreneurs! Therefore, it would seem as misguided today to speak of the decline of large organizations as it proved inaccurate twenty years earlier to speak of the fall of entrepreneurs. While smallness is *usually* more beautiful, bigness is simply a fact of life in a world where three billion people are increasingly linked by common technologies and markets.

It may be useful to think about this issue by way of analogy: Is the mega University of California going to give way to competition from hundreds of small colleges? Is the unitary government of France going to devolve all its power to the country's myriad *départements*? Is Boeing soon to give way to small-scale manufacturers of jumbo jets? While a reasonable answer to each of these questions is no, the most likely scenario is that the structures of giant universities, central governments, and colossal corporations will change to forms beyond our current ability to envision. Although we can't imagine exactly what these new structures will look like, it nonetheless seems reasonable to expect that almost all organizations that survive and thrive in the future will possess the best characteristics of today's big *and* small successes. That is why in so many well-led large organizations efforts are being made to overcome *dis*economies of scale by creating dozens of small, independent, manageable units.

The Small Mimic the Big

While the giants attempt to avoid extinction by imitating the behavior of fast-moving small companies, the parallel challenge for entrepreneurs in coming years is to build global markets by capturing the advantages of gargantuan firms. Fortunately, meeting this challenge will be facilitated by emerging, computer-based technologies of production and distribution. Newly developed manufacturing tools give small companies the advantages of mass production while at the same time allowing them to customize products economically. New telecommunications technologies provide access to distant and specialized markets that were formerly out of reach for all but giant firms with global distribution networks. Sophisticated data bases provide even the smallest companies with marketing information that just yesterday was affordable only to the largest. And all this technology is currently available. At present American fabric and apparel manufacturers are linked by computer to hundreds of retailers, thus giving increased purchasing power to the small firms and faster inventory information to the manufacturers, all of which factors permit U.S. companies to use technology to help overcome Asia's competitive wage advantage.

By fine-tuning the federal strategy by which the small, semiautonomous American states combined and cooperated in order to gain the advantages of a large nation, small businesses around the world are creating networks, partnerships, consortia, and federations—all designed to give them the functional equivalent of bigness. The best-known company pursuing a federal strategy is Benetton, where finance, R&D, design, purchasing, and planning are centralized, while the activities of manufacturing and retailing are dispersed. The company is a unique confederation of hundreds of small, manager-owned manufacturers and franchised retailers all linked together by computer to form the United States of Benetton. Like Benetton, such companies as Nike and The Limited also have learned that it is better to achieve the benefits of forward and backward integration through confederation rather than through acquisition.

Importantly, there is no single model of confederation. As Rosabeth Moss Kanter was the first to observe, companies around the world are "becoming PALs: Pooling, Allying and Linking" across corporate and national boundaries.[5] Small companies in particular are inventing all manner of joint ventures, subcontracting, franchising, R&D consortia, and strategic partnerships. These are taking the form of cooperation between customers and suppliers, between domestic and foreign entities, between large and small organizations—and even among competitors: After all, entrepreneurs are willing to do whatever it takes in order to combine the advantages of big and small. Some examples: Small record and book publishers (and film producers) use the services of large distributors to gain economies of scale in marketing; small airlines form consortia to buy jet aircraft from brokers in order to gain economies of scale in purchasing; small "hollow" corporations design furniture, contract to have it made in Third World countries, and then wholesale it to large department stores in Europe and America (or market their products themselves in stores-within-stores). As our colleague Jay Galbraith explains, the common thread in each of these examples is that small companies "buy the power of bigness"—that is, they have someone else provide the scale in marketing, purchasing, financing, or manufacturing that is uneconomical for the small company to attempt itself.

The federal form has applications not only for manufacturing and retailing but for service industries as well. American Airlines's SABRE system uses high technology to link the worldwide fortunes of numerous large and small competitors in the airline industry. In the United States nearly every service from real estate to plumbing has been successfully franchised, and international professional services firms like Arthur Andersen are in fact prime examples of the federal system. And Coca-Cola, with its global network of franchised bottlers and distributors, is the longest-standing—and most successful—example of the advantages of confederation.

Federalism as a Revitalization Strategy

Of more recent origin—and less conventional structure—is the confederation ABB, which employs more people around the world than live in the entire country of Iceland. Although some components of the company are over a hundred years old, ABB's CEO, Percy Barnevik, has demonstrated the validity of federalism as a strategy to revitalize old-line manufacturing firms for competition in today's world markets. Barnevik explains that ABB "is a company with no geographic center, no national ax to grind. We are a federation of national companies with a global communications center."[6] Barnevik is not worried by the contradictions that led Sloan to abandon federalism: "ABB is an organization with three internal contradictions. We want to be global and local, big and small, radically decentralized with centralized reporting and control."

The managerial secret that allows ABB to turn these contradictions into what Barnevik calls "real organizational advantage" is federalism with a vengeance. ABB's operations are divided "into nearly 1,200 companies, with an average of 200 employees. These companies are divided into 4,500 profit centers with an average of 50 employees." With only 100 professionals in its Zurich headquarters, the company is not unified by the efforts of an all-powerful central staff à la GM. Rather, this *non*centralized confederation of semiautonomous units is held together by a common vision of globalism, excellence, and clearly enunciated responsibilities for performance. What is the role of central headquarters? "To operate as lean as is humanly possible," says Barnevik. And the role of leadership? To give managers "well-defined sets of responsibilities, clear accountability, and maximum degrees of freedom to execute."

The Leadership Imperative

The sharpest image of the new federal leader that comes to mind is that of Coca-Cola's Roberto Goizueta at a recent meeting of the company's bottlers and distributors, where he was observed to implore those fiercely independent folks at least *three* times in *one* speech to "please paint your trucks red." How's that? In the year in which he earned some $80 million, the CEO of Coca-Cola had to plead with "his troops" to adhere to standards of corporate conformity? Clearly, something new is going on here. And that "something" is that leaders of federations don't think of their associates as troops—and the associates don't think of their leaders as generals.

Like ABB's, Coca-Cola's federalism is effective in a way that Sloan could never have imagined, because of a factor that emerged nearly three decades after the GM chief's death: a new concept of leadership. Sloan was a brilliant leader of GM, but therein lay the fatal flaw in his attempts to install federalism: Sloan was also the *only* leader at GM. In sharp contrast, the new leaders of the emerging federal corporations are *leaders of leaders* who, like Percy Barnevik and Roberto Goizueta, are willingly followed by other leaders who have subscribed to their "vision."

In the 1980s it became commonplace to regard the new leader as one who has the ability to generate a compelling, moving, and unifying vision. This means the ability to establish a climate and structure that give all members of the organization a clear sense of what they are doing and why. What has not been fully appreciated about "the vision thing" is that the purpose of a clearly communicated vision is to give meaning and alignment to the organization and thus to enhance the ability of *all* employees to make decisions and create change. The new leader does not make all decisions herself; rather, she removes the obstacles that prevent her followers from making effective decisions *themselves*. Thus, not only is the standard military leadership metaphor of generals and troops wrong; so is the classical peacetime metaphor of shepherds and sheep. The new leaders are no more shepherds than their followers are sheep. A more fitting metaphor is Schumacher's balloon man—now, a woman—who holds a fistful of strings attached to countless units, each tugging away because it is filled with the helium of entrepreneurial spirit.

Indeed, when we describe the emerging leadership relationship in today's federal organizations we come closest when we speak of *leaders of leaders.* In these organizations senior leaders are followed willingly by other leaders by virtue of the formers' vision, integrity, and courage (and not just by the organizational equivalent of a yank of the crook or the nipping of a sheepdog at the heels). Importantly, because people at *all* levels are leaders in their own right, there is little of the resistance to change that characterizes the middle ranks of most hierarchical organizations headed by a single commander in chief and staffed by layers of resentful sheep. In the emerging leadership relationship, it is far from easy for the outsider to identify *the* leader. As the chairman of Herman Miller, Inc., Max De Pree, explains, "The signs of outstanding leadership appear primarily among the followers. Are the followers reaching their potential? Are they learning? Serving? Do they achieve the desired results? Do they change with grace? Manage conflict?"[7] If so, the organization is blessed with an outstanding leader of leaders.

In the successful federal organization, a central—perhaps *the* central—task of the leader of leaders thus becomes the development of other leaders. At Dayton-Hudson Kenneth Macke spends about half of his time on the career development of the firm's top one hundred managers. With forty-five hundred employees in potential leadership positions, Percy Barnevik's job becomes one of creating the conditions in which all those people can succeed in their jobs. In effect, federalism provides a structural skeleton for the rhetorical goal of "empowerment." Thus, federalism does not obviate the need for leadership; instead, it focuses and redefines the task of the leader. The success of the current president of the European Commission of the EC, Jacques Delors, illustrates the necessity of federal leadership characterized by the provision of inspiring vision—coupled with the identification, nurturing, and development of future leaders empowered to carry out that vision.

Ultimately federalism may also pave the path toward more democratic organizations. When we ask, "Is democracy inevitable?" the answer is a more resounding and immediate "Yes" in federal systems. For as Jefferson and Madison recognized, democracy is more natural in smaller units and less wieldy in large, unitary states. Lest this federalism sound like softheaded, "touchy-feely" management, it is worth noting that George Will has called for a marked return to federalism in the American system of government: "That is the future—congressional ascendancy and vigorous federalism. We can live with that. The Founders said we should."

"A Pretty Good Alliance"

In essence federalism allows nations and corporations to have their organizational cake and eat it too. Given proper leadership, the New Federalism—whether in the guise of ABB or the EC—illustrates that it is possible to pursue innovation, self-governance, and autonomy, while at the same time enjoying the advantages of effective coordination, economies of scale, and the protection of cherished freedoms that only pluralism can provide. From a business perspective federalism erases the false "big-versus-small" dichotomy that has for too long preoccupied those engaged in debate about the essential traits needed for international competitiveness, much as it points the way toward variations on the theme of confederation that could lead to truly effective performance in the global economy.

Finally, we can imagine a time when corporations such as ABB—which are simultaneously global and deeply rooted in local cultures—serve as models for nations that aspire both to national self-expression and to survival in the world economy. These new confederations could resolve the Iceland Dilemma, and the only cost would be the loss of the jingoistic rhetoric of which national mottoes and state anthems have traditionally been composed. The slogans of the federations of the future probably won't be as stirring as the national slogans of the past. It is true that "My federation, a pretty good alliance" doesn't have the ring of "My country, right or wrong." But a world of overlapping and interwoven corporate and national federations would be a far better place in which to work and live.

Notes

1. James Madison, "Federalist Paper #10."

2. Alfred E. Sloan, *My Years with General Motors* (New York, Doubleday Anchor, 1972).

3. John Kenneth Galbraith, *The New Industrial State* (Boston: Houghton-Mifflin, 1967).

4. George Gilder, *Wealth and Poverty* (New York: Bantam Books, 1982).

5. Rosabeth Moss Kanter, *When Giants Learn to Dance* (New York: Simon and Schuster, 1989).

6. William Taylor, "The Logic of Global Business: An Interview with ABB's Percy Barnevik," *Harvard Business Review,* March–April 1991.

7. Max De Pree, *Leadership is an Art* (New York: Doubleday, 1989).

Index

Permission Acknowledgments

Is Democracy Inevitable?
Reprinted by permission of *Harvard Business Review.* "Democracy is Inevitable," by Warren G. Bennis and Philip Slater, Issue 90510 (September/October 1990). Copyright © 1990 by the President and Fellows of Harvard College. All rights reserved.

The Coming Death of Bureaucracy
Reprinted by permission from *Think Magazine,* an article entitled "The Coming Death of Bureaucracy," by Warren Bennis, copyright 1966 by International Business Machines Corporation.

Managing the Dream
Reprinted with permission from the May 1990 issue of *TRAINING Magazine,* "Managing the Dream," by Warren Bennis. Lakewood Publications, Minneapolis, Minnesota. All rights reserved.

On the Leading Edge of Change
Reprinted with permission from the April 1992 issue of *Executive Excellence,* "On the Leading Edge of Change," by Warren Bennis, copyright by Executive Excellence, Provo, Utah.

Searching for the Perfect University President
Reprinted with permission from *The Atlantic Monthly,* "Searching for the 'Perfect' University President," by Warren Bennis, copyright 1971 by The Atlantic Monthly Company. All rights reserved.

When to Resign
Reprinted with permission from *Esquire,* "When to Resign," by Warren Bennis, copyright 1972 by Esquire and The Hearst Corporation.

Followership
From the December 31, 1989 issue of *The New York Times,* "Forum/The Dilemma at the Top; Followers Make Good Leaders Good," by Warren Bennis, Copyright © 1989 by The New York Times Company. Reprinted by permission.

Ethics Aren't Optional
Reprinted with permission from the April 1988 issue of *Executive Excellence*, "Ethics Aren't Optional," by Warren Bennis, copyright by Executive Excellence, Provo, Utah.

Change: The New Metaphysics
Reprinted with permission from the November 1990 issue of *Executive Excellence*, "Change: The New Metaphysics," by Warren Bennis, copyright by Executive Excellence, Provo, Utah.

Meet Me in Macy's Window
Reprinted with permission from the October 1975 issue of *Harvard Magazine*, "Meet Me in Macy's Window," by Warren Bennis, copyright © 1975 by Harvard Magazine.

Corporate Boards
"The Crisis of Corporate Boards," (November 1978) and "RX for Corporate Boards" (December/January 1979), by Warren Bennis, reprinted with permission from *Technology Review*, copyright 1978 and 1979.

Information Overload Anxiety
"Information Overload Anxiety," (July/August 1979) by Warren Bennis, reprinted with permission from *Technology Review*, copyright 1979.

Our Federalist Future
"Our Federalist Future," by Warren Bennis and James O'Toole, copyright 1992 by The Regents of the University of California. Reprinted from the *California Management Review*, Vol. 34, No. 4. By permission of The Regents.